RANDOM GENERATION OF TREES

RANDOM GENERATION OF TREES

Random Generators in Computer Science

by

Laurent Alonso

and

René Schott

CRIN, Université Henri Poincaré

KLUWER ACADEMIC PUBLISHERS

BOSTON / DORDRECHT / LONDON

A C.I.P. Catalogue record for this book is available from the Library of Congress.

ISBN 978-1-4419-5150-2

Published by Kluwer Academic Publishers,
P.O. Box 17, 3300 AA Dordrecht, The Netherlands.

Kluwer Academic Publishers incorporates
the publishing programmes of
D. Reidel, Martinus Nijhoff, Dr W. Junk and MTP Press.

Sold and distributed in the U.S.A. and Canada
by Kluwer Academic Publishers,
101 Philip Drive, Norwell, MA 02061, U.S.A.

In all other countries, sold and distributed
by Kluwer Academic Publishers Group,
P.O. Box 322, 3300 AH Dordrecht, The Netherlands.

Printed on acid-free paper

CONTENTS

Le Chêne et le Roseau

Par Jean de La Fontaine

Un chêne un jour dit au roseau :
≪ Vous avez bien sujet d'accuser la nature ;
Un roitelet pour vous est un pesant fardeau ;
Le moindre vent qui d'aventure
Fait rider la face de l'eau ;
Vous oblige à baisser la tête,
Cependant que mon front, au Caucase pareil,
Non content d'arrêter les rayons du soleil,
Brave l'effort de la tempête.
Tout vous est aquilon, tout me semble zéphyr.
Encore si vous naissiez à l'abri du feuillage
Dont je couvre le voisinage,
Vous n'auriez pas tant à souffrir :
Je vous défendrais de l'orage ;
Mais vous naissez le plus souvent
Sur les humides bords des royaumes du vent.
La nature envers vous me semble bien injuste.
-Votre compassion, lui répondit l'arbuste,
Part d'un bon naturel ; mais quittez ce souci :
Les vents me sont moins qu'à vous redoutables ;
Je plie et ne romps pas. Vous avez jusqu'ici
Contre leurs coups épouvantables
Résisté sans courber le dos ;
Mais attendons la fin. ≫ Comme il disait ces mots,
Du bout de l'horizon accourt avec furie
Le plus terrible des enfants
Que le Nord eût portés jusque-là dans ses flancs.
L'arbre tient bon ; le roseau plie.
Le vent redouble ses efforts,
Et fait si bien qu'il déracine
Celui de qui la tête au ciel était voisine,
et dont les pieds touchaient à l'empire des morts.

PREFACE

This text is about a field on the crossroads between computer science, combinatorics and probability theory. Computer scientists need random generators for performance analysis, simulation, image synthesis, etc ... In this context random generation of trees is of particular interest. The algorithms presented here are efficient and (often) easy to code. Some aspects of Horton-Strahler numbers, programs written in C and pictures are presented in the appendices. The complexity analysis is done rigorously both in the worst and average cases.
This book is intended for students in computer science and/or applied mathematics and for researchers interested in random generation. There is didactical material for the former group and there are advanced technical sections for the latter group.
We would like to acknowledge our colleagues at the University of Nancy-the CRIN, INRIA Lorraine-whose support and encouragement is truly appreciated. Especially, the authors are grateful to J.L.Rémy for early discussions relative to the random generation of trees. We would like as well to acknowledge our colleagues from Bordeaux (A.Denise, Samaj Lareida, J.G.Penaud, X.G.Viennot and many others), Paris (J.Berstel, D.Gouyou-Beauchamps and M.Pocchiola), Urbana-Champaign (N.Dershowitz, C.L.Liu, E.M.Reingold) for pertinent comments which contributed to improve the presentation and the contents of this book. P.Feinsilver and U.Franz helped us with the intricacies of the English language. Finally we express gratitude to our families and to all our friends. Their patience and encouragement made this project possible.

1

INTRODUCTION

Trees are a combinatorial structure that has many important applications in computer science. They are highly appreciated as elementary tools. They are used, for example, to sort and search data (binary trees, AVL trees [2], 2-3 trees, red-black trees [11], ...), to code images (quad-trees), and to represent mathematical expressions in a form that can be manipulated more easily :

Example :

The tree

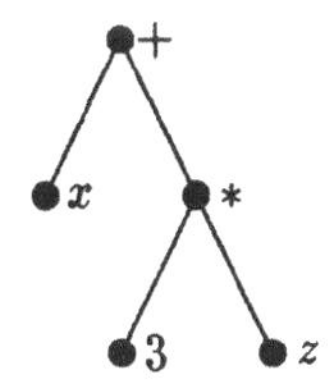

is a correct representation of the expression : $x + 3 * z$.

Generally speaking, they are a good instrument to save structures that can be described recursively (tuples in data bases, ...), or that appear when faster algorithms, using methods of the type "divide and conquer" or more heuristic methods (in-depth search, "min-max" algorithm), are employed.

Trees are also very useful for analyzing the behavior of a large number of algorithms. In the first place, they allow visualizing the calls of a recursive algorithm :

For example :

If we define the Fibonacci function by

fib (0) = 1,
fib (1) = 1,
fib (n+2) = fib (n+1) + fib (n),

then the calls performed to calculate $fib(4)$ can be represented by the tree :

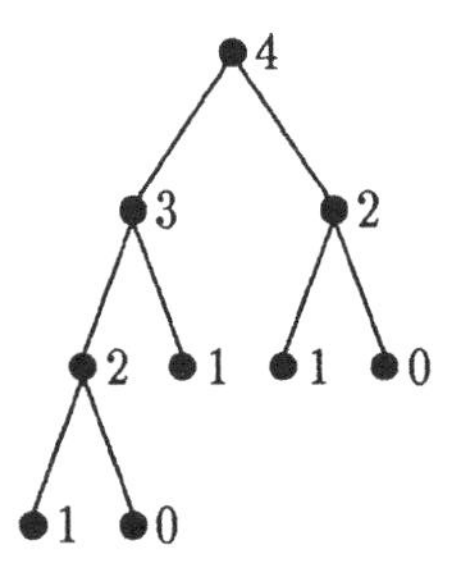

($fib(4)$ creates two calls, the first is $fib(3)$, which, in turn, ...).

We obtain a tree that has $fib(4)$ leaves and $fib(4) - 1$ inner vertices. Therefore $fib(4) - 1$ additions had to be performed to calculate $fib(4)$.

They also appear as structures that are homeomorphic to other structures which are used in certain analyses. In this way the language of brackets, often called Dyck language, is closely related to binary trees. Similarly, the Motzkin words have a close connection to the unary-binary trees ([51]), ...

We will study here the problem of generating randomly tree structures.

Two different viewpoints are possible.

- In the first, we choose a set of trees $\mathcal{E}$ (for example the binary trees of size 131, the ternary trees of height 20), and look for an algorithm that generates a tree of this set with uniform probability (i.e. $\frac{1}{card(\mathcal{E})}$).
- On the other hand, in the second approach, we also start by selecting a set $\mathcal{E}$ of trees, but look for an algorithm that creates all trees of $\mathcal{E}$ one after the other. Results in this area have been obtained for k-ary trees of size n [23],

for arbitrary trees with n inner vertices and m leaves [40], for binary trees of height h [35], for B-trees [31], for generating binary trees on computers with multi-processors [4], ...

We will consider the first problem here.

There exist many ways to generate randomly a tree. If we try to classify the different kinds of algorithms, we get the following classification :

nethods by growing : We present in Chapter 3 some of these methods. They consist in building a small tree then in making it grow. The obtained algorithms are very nice and efficient. However, there exist, to our knowledge, very few algorithms which use this approach.

nethods using 1-1 correspondence : Chapter 3 contains some of the algorithms using this method and Chapters 4, 5 and 6 present some 1-1 correspondences between some trees and some words and forests that can be used to generate some trees. These methods consist in finding a class of combinatorial objects which can be easily generated and which is in 1-1 correspondence with the trees that we want to build. The obtained algorithms are efficient, easy to code and allow the generation of many kinds of trees which can be enumerated by simple formulas.

When this is not the case, the two following methods are often useful.

nethods by rejection: We present in Chapter 7 some algorithms which use it. These methods consist in generating an element of a set $\mathcal{E}'$ ($\mathcal{E}'$ contains $\mathcal{E}$) then in keeping this element if it is in the set $\mathcal{E}$ and in going back to the beginning if not. These methods give efficient algorithms but unfortunately we know only few algorithms which use it.

method by recursion : We present in Chapters 8 and 9 some algorithms which use it and some tricks to improve their real complexity. This method uses the recursive definition of a tree to find a generation algorithm. This approach is really powerful, however it gives often algorithms whose complexity is greater than when we can use the preceding method. A more complete description of the power of these methods is in [27].

Chapter 10 presents a parallel algorithm for the generation of a permutation and shows how to generate a Dyck (respectively Motzkin) word and corresponding left factors on a parallel machine. As we will see, binary (respectively unary-binary) trees are in 1-1 correspondence with Dyck (respectively Motzkin) words. The same technique applies also to the generation of words which are in bijection with trees split into patterns as defined in Chapter 4.

Using random generators is a good way to obtain images of trees that are close to reality. D. Arquès, G. Eyrolles, N, Janey, G. Viennot (cf. [1] and [24]) have suggested several methods to get an image of a tree. Starting with a binary tree they use the Strahler number of its vertices. This number is recursively defined as

- 1 for a leaf,
- $l+1$ for a vertex that has two children, if both children have Strahler number equal to l,
- $max(l_1, l_2)$ for a vertex with two children with different Strahler numbers. The numbers l_1 and l_2 denote the Strahler numbers of the children ($l_1 \neq l_2$).

Applying these methods to different kinds of trees we obtained the drawings that illustrate this book.

2

NOTATIONS

Abstract

In this chapter we give a set of notations and definitions that will be used in this book.

2.1 SOME MATHEMATICAL NOTATIONS

We will use in the next chapters

- $\lfloor x \rfloor$, floor of x : $max_{n\, integer, n \leq x} n$,
- $\lceil x \rceil$, ceiling of x : $min_{n\, integer, n \geq x} n$,
- $[\![a, b]\!]$: the set of the integers n such that $a \leq n \leq b$,
- $n!$, factorial of n : $1.2.\cdots.n$,
- $\binom{n}{k}$, the number of different ways to choose k elements between n : $\frac{n!}{k!(n-k)!}$,
- $\binom{n}{k_1, k_2, \cdots, k_p} = \frac{n!}{k_1!k_2!\cdots k_p!(n-k_1-k_2\cdots-k_p)!}$: the number of different ways to choose k_1 elements 1, ..., k_p elements p between n : $\frac{n!}{k_1!k_2!\cdots k_p!(n-k_1-k_2\cdots-k_p)!}$,
- $Log(x)$, natural logarithm of x,

- $ln(x)$, logarithm in base 2 of x,
- $log_p(x)$, logarithm in base p of x : $\frac{Log(x)}{Log(p)}$,
- $Ln_p(x)$, iterative logarithm defined by
  - when $p = 1$: $Ln_1(x) = Ln(x) = max(1, Log(x))$,
  - when $p > 1$: $Ln_p(x) = Ln(Ln_{p-1}(x))$,
- $a(n) \sim b(n)$: $\lim_{n \mapsto \infty} \frac{a(n)}{b(n)} = 1$,

2.2 DEFINITIONS OF TREES

First, we recall the definition of a tree :

Definition 1 ■ *A tree on the finite set $S \neq \emptyset$ is a pair $(r, (T_1, T_2, \ldots, T_k))$ (with $k \geq 0$) such that*

- *r is an element of S,*
- *T_1, ..., T_k are some trees defined on the sets S_1, ..., S_k,*
- *the set $\{r\}$, S_1, ..., S_k form a partition of S (ie. $S = \{r\} \cup S_1 \cup S_2 \cdots \cup S_k$ and $\{r\} \cap S_i = S_i \cap S_j = \emptyset$ for each i and j in $[\![1, k]\!]$ and $j \neq i$).*

We call nodes or vertices the elements of S, root of T the node r. We say that there exist k edges that connect the node r to the root of the trees T_1, ..., T_k and that the roots of T_1, ..., T_k are the children of the node r.

Furthermore, we consider that two trees T and T' are the same if there exists an isomorphism from S to S' which transforms T into T'.

- *A forest is a list of trees $(T_1, T_2, \ldots, T_p)$.*

Below, we fix some more definitions. We denote:

- by arity of a node the number of its children,
- by leaf a node with no child,
- by inner node a node with at least one child,

- by unary node a node with one child,
- by binary node a node with two children,
- by ternary node a node with three children,
- by k-ary node a node that has k children,
- by binary (resp. ternary, k-ary) tree a tree whose inner nodes are binary (resp. ternary, k-ary) nodes,
- by unary-binary tree a tree whose inner nodes are either unary or binary nodes,
- by size of T : $_{\#}T$ the number of nodes that are in T (ie. $card(S)$),
- by path which connects the nodes n_1 and n_p a list $(n_1, n_2, \cdots, n_p)$ such that there exists an edge which connects the nodes n_i and n_{i+1} for each i in $[\![1, p-1]\!]$,
- by height of a node v in T the minimal number of distinct nodes of any path connecting v to the root of T,
- by height of T the maximum value of the height of the nodes of T.

Example :

Below are three trees :

each node is represented by the symbol ● and each edge by a line segment which connects two nodes ● . Furthermore, the label r indicates the root of each tree.

The size of these trees is 5, 1, 11 from the left to the right and their height is 3, 1, 4.

Now we display a forest of three binary trees :

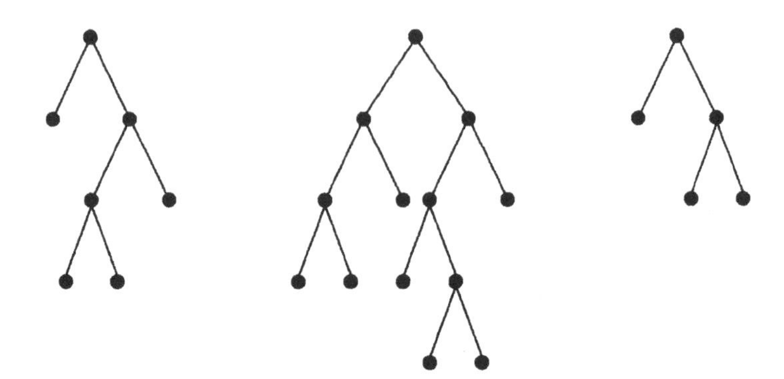

a ternary tree :

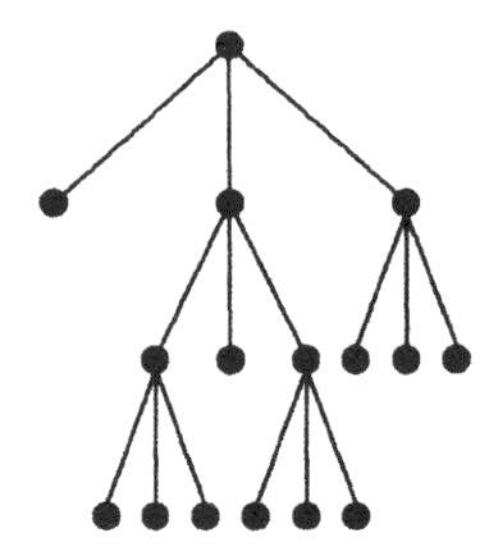

and a unary-binary tree

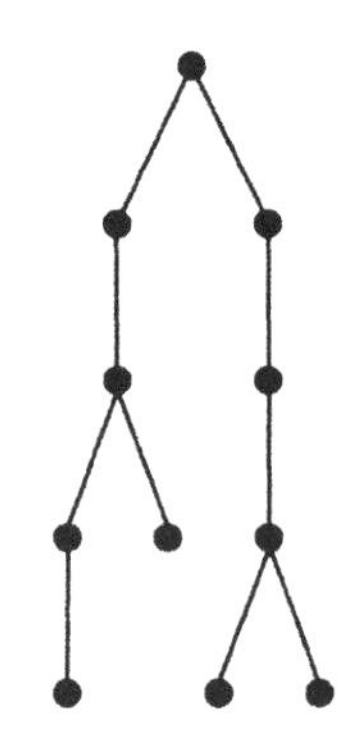

Now we define two different tree traversals. The corresponding two recursive procedures are as follows :

prefix(x)
if x is the root of a subtree
    print (x)

```
    denote by x_1, ..., x_p the children of x,
    i = 1
    while i ≤ p
          prefix (x_i)
```

and

```
postfix(x)
if x is the root of a subtree
    denote by x_1, ..., x_p the children of x,
    i = 1
    while i ≤ p
          postfix (x_i)
    print (x)
```

We can now define the prefix (respectively postfix) position of a node $x \in T$ as its position in the sequence $prefix(T)$ (resp. $postfix(T)$).

Example :

Consider the tree T defined by

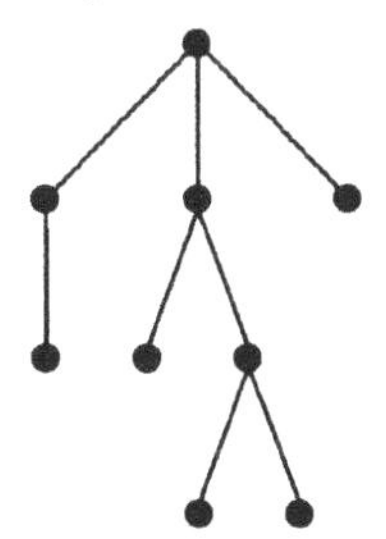

when we label the nodes of T with their prefix positions, we get:

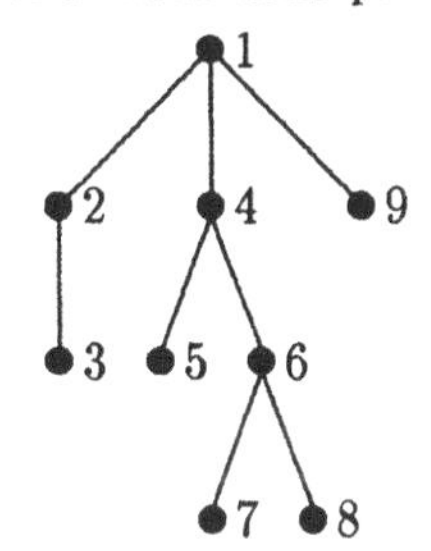

and with their postfix positions :

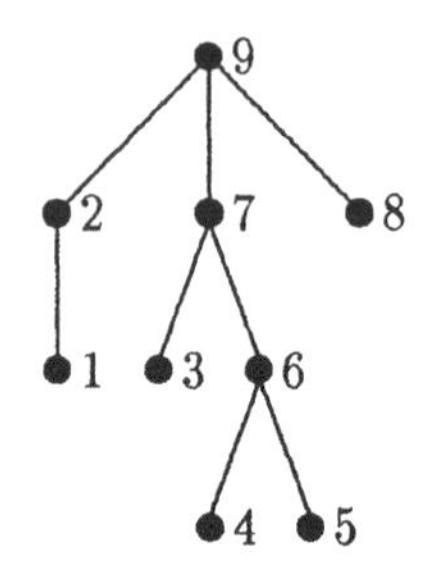

We can now define another kind of trees (called arborescences) by

Definition 2 *An arborescence on the finite set* $S \neq \emptyset$ *is a pair* $(r, \{T_1, \ldots, T_k\})$ *(with* $k \geq 0$*) such that*

- r *is an element of* S,
- $T_1, \ldots, T_k$ *are some arborescences defined on the sets* $S_1, \ldots, S_k$,
- *the set* $\{r\}$, $S_1, \ldots, S_k$ *form a partition of* S.

We call nodes the elements of S, *root of* T *the node* r *and we say that there exist* k *edges that connect the node* r *to the root of the arborescences* $T_1, \ldots, T_k$.

Furthermore, we consider that two arborescences T *and* T' *are the same if there exists an isomorphism from* S *to* S' *which transforms* T *into* T'.

Remark :

As opposed to the trees, the children of a node in an arborescence are not ordered. Therefore there will be many pictures which represent the same arborescence.

For instance the following pictures represent the same arborescence :

We use the same terms to speak about arborescences and trees, we only replace in the definition of these terms the name tree by arborescence each time it appears.

2.3 OTHER DEFINITIONS

We give first some definitions about words

Definition 3 *Denote by w a word then:*

- *t is a factor of w if there exist two words u and v such that $w = utv$. If $u = \epsilon$ (ie. the word with no letter) then t is a left factor (or prefix) of w. If $v = \epsilon$, t is a right factor(or suffix) of w.*
- *We denote by $|w|$ the number of letters of w and by $|w|_x$ the number of letters x in w.*
- *If $|w| \geq i$, we denote by w_i the prefix of w with i letters.*

In the following we consider only words with a finite number of letters.

Now we introduce two kinds of words : Dyck words and Motzkin words.

Definition 4 *Let w be a word formed with the letters x and y, we will say that*

- *w is a Dyck word if w is either an empty word or a word $xuyv$ where u and v are some Dyck words. If we denote by D the language of the Dyck words we obtain :*
$$D = \epsilon + xDyD$$
- *w is a Dyck left factor if and only if w is a left factor of a Dyck word,*
- *w is a 1-dominated sequence if and only if there exists a Dyck left factor u such that $w = xu$.*
- *if l is the ith letter of w, we will denote by height(l) the number $|w_i|_x - |w_i|_y$.*

Example :

The word $w = xxyyxy$ is a Dyck word and $w = xxyyx$ is a Dyck left factor.

If $w = xyyxyyyyx$ then the height of the letters of w are 1, 0, −1, 0, −1, −2, −3, −4, −3 from the left to the right.

Definition 5 *Let w be a word formed with the letters x, y, a then*

- *w is a Motzkin word if and only if*
  - *w is the empty word,*
  - *or $w = av$ where v is a Motzkin word,*
  - *or $w = xuyv$ where u and v are two Motzkin words.*

  If we denote by M the language associated to the Motzkin words, we get

$$M = \epsilon + aM + xMyM,$$

- *w is a Motzkin left factor if w is a left factor of a Motzkin word.*

Example :

The word $axaxyya$ is a Motzkin word and $axaxy$ is a Motzkin left factor.

Finally, we define the ordinary generating function by

Definition 6 *Let $\mathcal{E}$ be a set and denote by t_n the number of elements of $\mathcal{E}$ of size n, the ordinary generating function f of the set $\mathcal{E}$ is the formal function defined by :*

$$f(z) = t_0 + t_1 z + t_2 z^2 + \cdots = \sum_{i=0}^{\infty} t_i z^i.$$

Example :

If $\mathcal{E}$ is the set of binary trees, we get (see [51]):

$$f(z) = z + z^3 + 2z^5 + \cdots = \frac{1 - \sqrt{1 - 4z^2}}{2z}.$$

Remark :

These functions are sometimes really useful to determine the asymptotic behavior of the numbers t_i[51].

3

GENERATION OF SIMPLE KINDS OF TREES

Abstract

In this chapter we present three algorithms that can generate three kinds of trees in linear time: binary trees, arbitrary trees, increasing trees, and Cayley arborescences.

3.1 GENERATION OF BINARY TREES: RÉMY'S METHOD

We will see in this paragraph how to generate binary trees with n internal vertices. First, we will show that this is equivalent to generating binary trees with n inner vertices and labelled leaves. We then show how to generate those new trees.

3.1.1 Binary trees and binary trees with labelled leaves

We have the following classical theorem which allows us to determine the size of a binary tree as a function of its number of inner vertices.

Property 1 *A binary tree with n inner vertices has $n + 1$ leaves.*

Proof. The proof is done by induction on n.

The property is true for $n = 0$; indeed, there is only one binary tree without inner vertices, the tree with only one vertex. This tree has one leaf.

We choose a positive integer N and assume that the proposition is true for all integers n which are strictly smaller than N.

Let T be a binary tree with N inner vertices. Since $N \geq 1$, T has a root r which has two subtrees, T_1 and T_2, as children.

Let n_1 and n_2 denote the number of inner vertices of T_1 and T_2. Since every inner vertex of T is either the root of the tree T or contained in one of the subtrees T_1 and T_2, we have $N = 1 + n_1 + n_2$.

But as n_1 and n_2 are smaller or equal to $N - 1$, we can apply the hypothesis of the induction to the subtrees T_1 and T_2. The subtree T_1 has therefore $n_1 + 1$ leaves and T_2 has $n_2 + 1$ leaves.

If we observe that every leaf of T is contained in either T_1 or T_2, we find that the number of leaves of T is $n_1 + 1 + n_2 + 1 = N + 1$. □

We can now proceed with the definition of binary trees with labelled leaves.

Definition 7 *A binary tree with labelled leaves is a binary tree whose leaves are labelled by numbers between* 1 *and the number of leaves of that tree. The numbers on the leaves are pairwise different.*

Theorem 1 *A binary tree of size* $2n+1$ *can be transformed into a binary tree with labelled leaves by attaching numbers to all its leaves in* $(n + 1)!$ *ways.*

Proof. We just saw that a binary tree with n inner vertices has $n + 1$ leaves. Thus a binary tree with $2n + 1$ vertices has n inner vertices and $n + 1$ leaves.

Now we have to label each of the $n + 1$ leaves of our binary tree by a number between 1 and $n + 1$, so that the labels are all distinct.

This problem is equivalent to choosing the leaf that gets the number $n + 1$ and then distributing the numbers from 1 to n on the remaining n leaves.

We therefore have $(n + 1)n(n - 1) \cdots 1 = (n + 1)!$ possibilities. □

Since each binary tree with n internal vertices corresponds to $(n+1)!$ binary trees with labelled leaves, we can generate a binary tree of size $2n+1$ randomly by first generating a binary tree with labelled leaves with n internal vertices and then removing the labels from its leaves.

3.1.2 Generation of binary trees with labelled leaves

Here we will present Rémy's recursive algorithm for constructing a binary tree with labelled leaves of size $2n+1$. To do this we assume that we know how to generate a tree T of size $2n-1$ and we add to this tree an inner vertex and the leaf that carries the label $n+1$.

We begin by choosing a vertex v of T with uniform probability (each vertex of T is chosen with probability $\frac{1}{2n-1}$). Then we choose the direction "right" or "left" with probability $\frac{1}{2}$. We replace the subtree v of T by a root which has the leaf with the label $n+1$ and the old subtree of the vertex v of T as children. The leaf will be placed in the direction that was chosen with probability $\frac{1}{2}$.

Example :

Taking the tree presented below and choosing the vertex marked by a circle and the direction "right", we get :

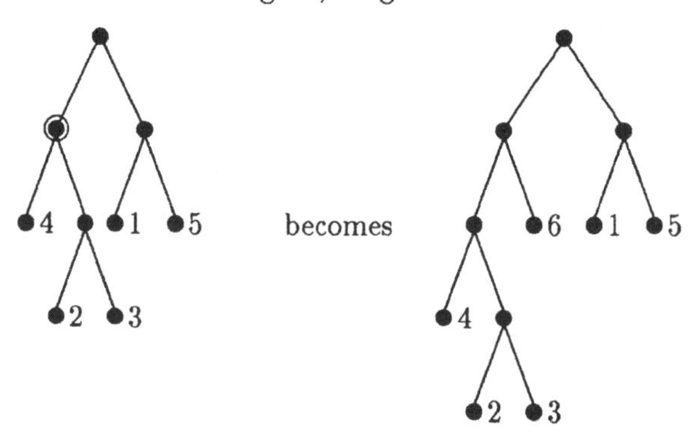

It is obvious that if we start with the tree consisting of one single leaf, we will get a binary tree with labelled leaves of size $2n+1$ in n steps. It remains to show that each binary tree with labelled leaves of size $2n+1$ has the same probability to be chosen. The content of the following theorem expresses this.

Theorem 2 *This algorithm generates all binary trees with labelled leaves of size $2n+1$ with probability $\frac{n!}{2n!}$.*

Proof. We will proceed by induction.

The theorem is true for $n = 0$; the algorithm yields the tree consisting of one single leaf with label 1 with probability $1 = \frac{0!}{0!}$.

Suppose now that the algorithm constructs the trees of size $2n+1$ with probability $\frac{n!}{2n!}$, and let T be a binary tree with labelled leaves of size $2n+3$.

We denote now by T' the tree obtained from T by removing the leaf u labelled by $n+2$ and replacing the parent w of u by the second child v of w as described below :

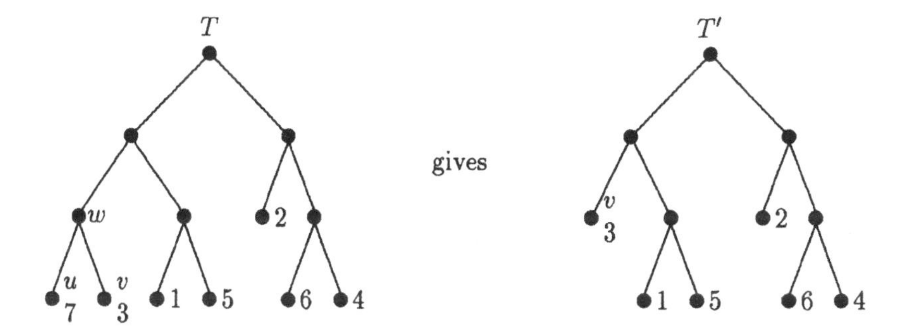

The only way to generate the tree T is to construct the tree T' and to choose the vertex v and the direction corresponding to the leaf u in T. This shows us that the probability $p(T)$ of generating the tree T is equal to

$$\frac{p(T')}{2(2n+1)}.$$

Applying the hypothesis of the induction we can conclude the following :

$$p(T) = \frac{n!}{2n!} \frac{1}{2(2n+1)} = \frac{(n+1)!}{(2n+2)!}.$$

□

3.1.3 Generation of arbitrary trees

We can use the preceding algorithm to generate arbitrary trees with n vertices.

Indeed, a bijective mapping exists that transforms a binary tree with n inner vertices into an arbitrary tree with $n+1$ vertices.

This bijection can be given as follows :

- we start by taking a binary tree with n inner vertices

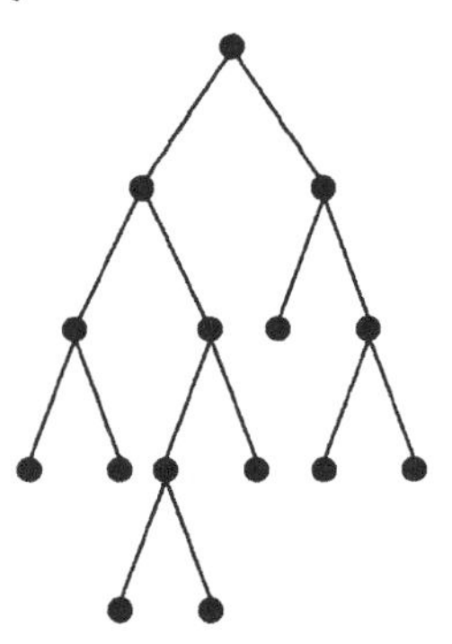

- we remove all leaves and represent the edges connecting a vertex to its left child by vertical edges and the edges connecting a vertex to its right child by horizontal edges

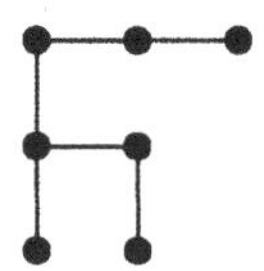

- we add now one vertex at the top of the diagram and connect this vertex to the root of this new tree by a vertical edge

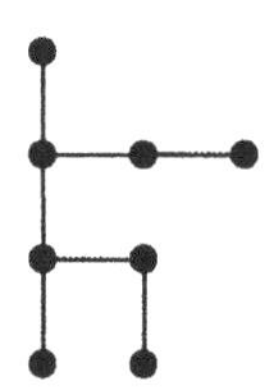

- we remove now all horizontal edges and connect each vertex to the first of its ancestors (i.e. the parent of its leftmost sibling) which is located on a higher level

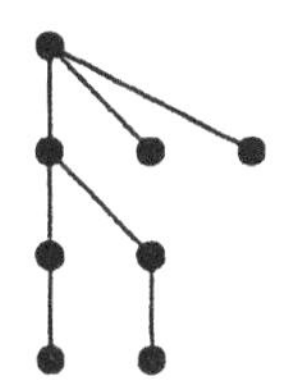

This transformation can be coded linearly, and thus gives us a method to generate arbitrary trees : first we generate a binary tree of size $n-1$ and then we transform this tree into an arbitrary tree of size n using the transformation described above. In the following chapter we will see another method to generate this type of tree with linear complexity.

3.2 GENERATION OF INCREASING TREES

Here we are going to show how to generate increasing trees : a kind of trees with labels. The algorithms are mainly based on bijective mappings.

3.2.1 Increasing binary trees

Definition 8 *An increasing binary tree on an ordered finite set S is*

- *the one node tree* $\bullet$*, if S is empty,*
- *A triple (r, T_1, T_2) where*
  - *r is the minimal element of S,*
  - *T_1 and T_2 are two increasing binary trees on some sets S_1 and S_2,*
  - *the sets $\{r\}$, S_1, and S_2 form a partition of S.*

Example :

An increasing binary tree on the set $[\![1, 9]\!]$:

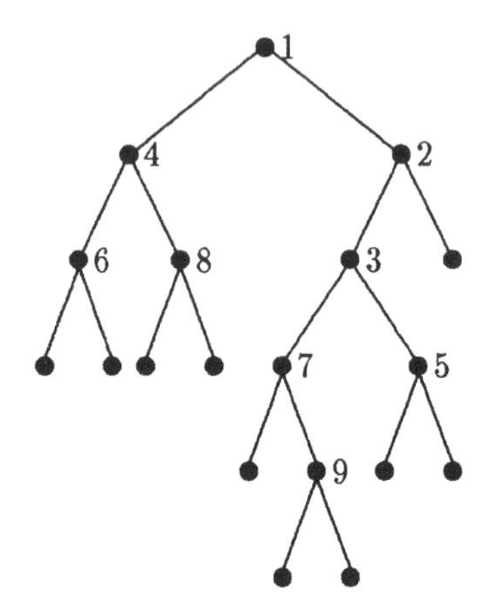

If we move down along the tree starting from the root, then the numbers we encounter always form an increasing sequence ; this justifies the term "increasing" in the name for this kind of tree.

Construction

Proposition 1 *There exists a bijection between the increasing binary trees on the set* $[\![1, n]\!]$ *with one leaf marked and the increasing binary trees on the set* $[\![1, n+1]\!]$.

Proof. Choose an arbitrary increasing binary tree T on the set $[\![1, n]\!]$ and choose one leaf v out of its $n+1$ leaves. We can transform this tree T into an increasing binary tree on the set $[\![1, n+1]\!]$ by replacing the leaf v by a vertex carrying the label $n+1$. This vertex has, as children, two unlabelled leaves as shown below :

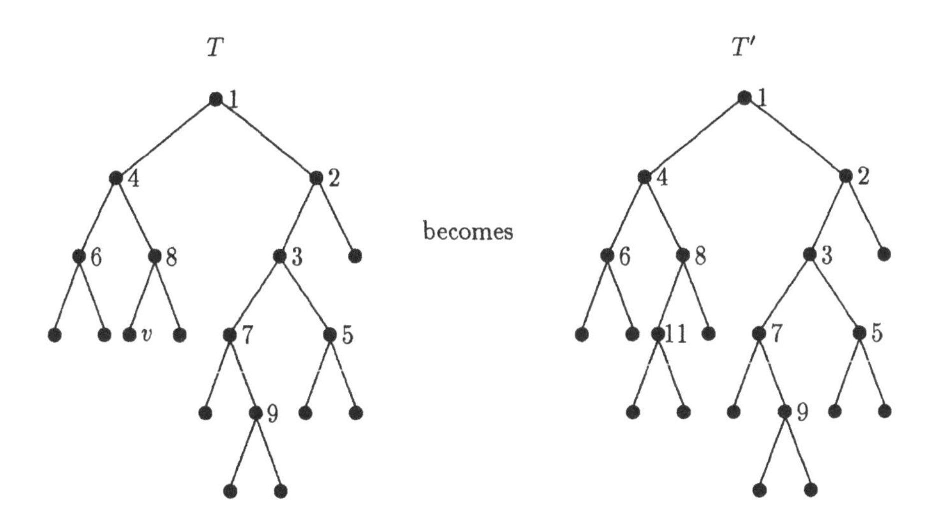

This transformation is bijective. □

We can now use this algorithm to find a recursive algorithm. In order to construct an increasing binary tree on the set $[\![1, n]\!]$, it is sufficient to

- construct a random increasing binary tree T on $[\![1, n-1]\!]$,
- choose one of the n leaves of this tree,
- replace this leaf by a vertex which carries the label n and has two unlabelled leaves as children.

3.2.2 Increasing trees

These trees resemble very much the increasing binary trees defined in the preceding paragraph. They can also be generated using similar techniques. First we give their definition :

Definition 9 *An increasing tree on an ordered finite non-empty set S is a $p+1$-tuple $(r, T_1, \ldots, T_p)$ (with $p \geq 0$) where*

- *r is the minimal element of the set S,*

- *T_1, ..., T_p are increasing trees on the sets S_1, ..., S_p,*
- *$\{r\}$, S_1, ..., S_p form a partition of S.*

Example :

An example of an increasing tree on the set $[\![1, 9]\!]$:

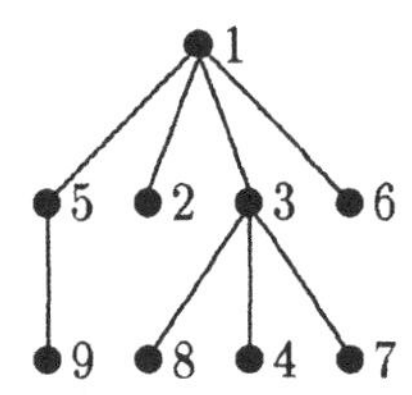

As in the previous case, we will construct the trees by letting them grow. If we take an increasing tree T on the set $[\![1, n]\!]$, then there are $2n - 1$ ways in which the vertex with the label $n + 1$ can be added to T to give an increasing tree on the set $[\![1, n + 1]\!]$:

- we can add the new vertex as the right child of one of the n vertices of T. If we choose the vertex marked by a circle, the tree

  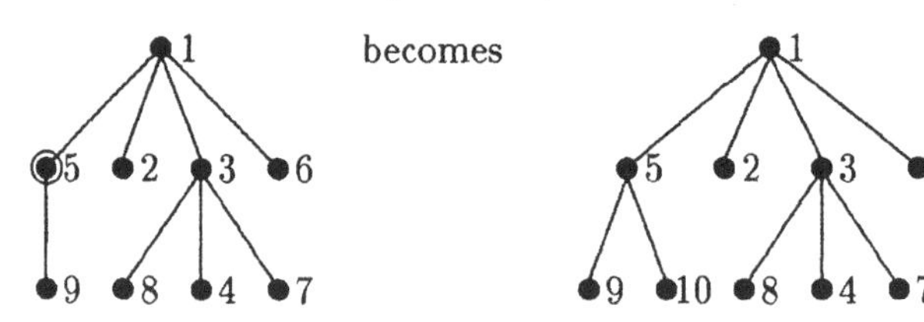

  There are n ways to do this.

- we can also choose one of the $n - 1$ vertices which are not at the root of our tree and add the vertex with label $n + 1$ as left sibling of v. Choosing the vertex with the circle, the tree

  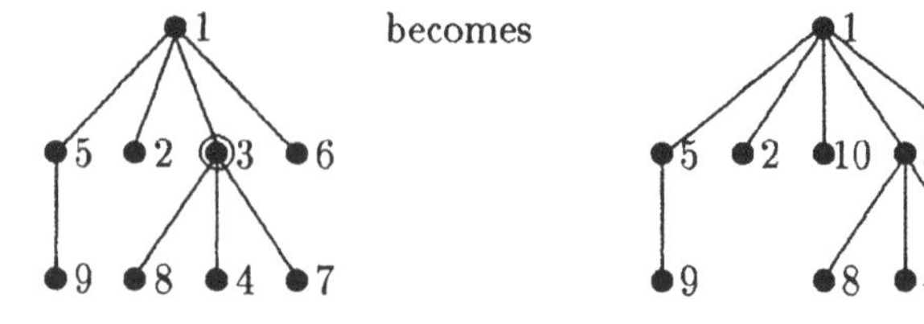

There are $n - 1$ ways to do this.

The algorithm for generating randomly an increasing tree on the set $[\![1, n]\!]$ consists therefore in constructing an increasing tree T on the set $[\![1, n-1]\!]$ and choosing a vertex or edge of T.

- If we choose one of the n vertices of T, then we attach the vertex carrying the label n as right child to this vertex,
- otherwise we add an edge to the left of the chosen edge and attach the vertex labelled by the integer n to this new edge.

3.2.3 Increasing arborescences

We will now show how to generate increasing arborescences by employing a bijection between this kind of tree and increasing binary trees.

Definition 10 *An increasing arborescence on an ordered finite set S is a pair $(r, \{T_1, \ldots, T_p\})$ where*

- *r is the minimal element of the set S,*
- *T_1, ..., T_p are increasing arborescences on the sets S_1, ..., S_p,*
- *$\{r\}$, S_1, ..., S_p form a partition of S.*

Remark :

Since the children of the vertices are not ordered, we have to assign them an arbitrary order to draw them.

Just like binary trees and arbitrary trees, there also exists a bijection between increasing arborescences and increasing binary trees.

This bijection consists of taking an increasing binary tree on the set $[\![2, n]\!]$

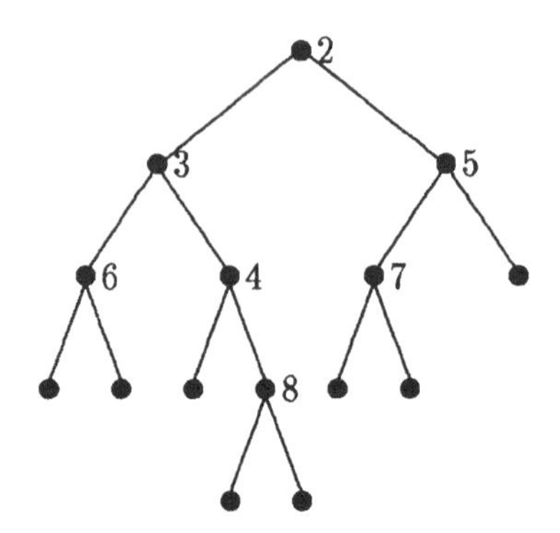

and then removing the unlabelled leaves, adding the vertex 1 and replacing the edges to the left by vertical edges and the edges to the right by horizontal edges.

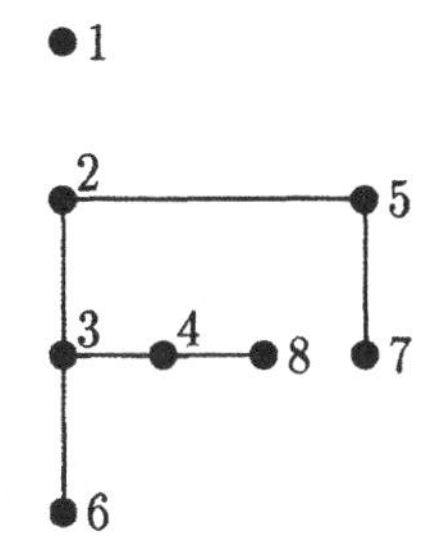

Next, we remove the horizontal edges and connect the vertices which have become unattached to the vertex on the upper level (ie. the parent of its leftmost sibling)

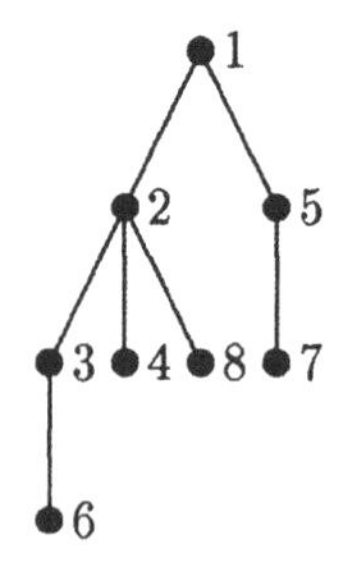

So, in order to generate an increasing arborescence with n nodes, first we generate an increasing tree on the set $[\![2, n]\!]$ then we use the 1-1 correspondence between these trees and the increasing arborescences to transform the obtained tree into an increasing arborescence with n nodes.

3.3 GENERATION OF CAYLEY ARBORESCENCES

3.3.1 Definition

We are now going to study another kind of labelled arborescences. These arborescences are defined as follows:

Definition 11 *A Cayley arborescence is a sequence of vertices labelled by the integers from* 1 *to* n *and connected by edges two by two. This structure has the following property : a unique path which connects any two arbitrary vertices exists.*

Example :

A Cayley arborescence of size 10 :

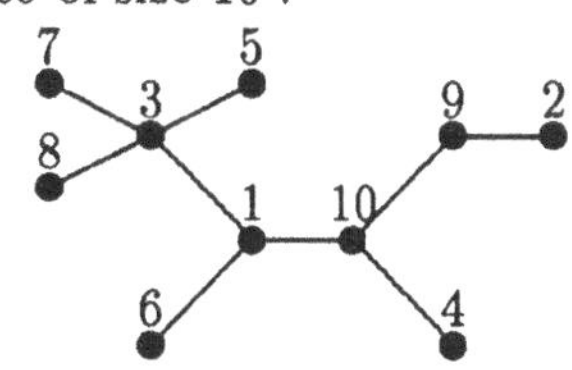

3.3.2 Transformation of a Cayley tree of size n into an element of $[\,1, n\,]^{n-2}$

The Cayley arborescences can be coded in the form of sequences of $n-2$ numbers of the interval $[\,1, n]$. The coding consists in recursively removing the smallest leaf of the Cayley arborescence and adding the label of the vertex that was attached to this leaf.

We treat the tree from the previous example in this manner : the smallest leaf has the label 2, and it is connected to the vertex 9. Therefore the sequence starts with the number 9. We are left with the task to code the following tree :

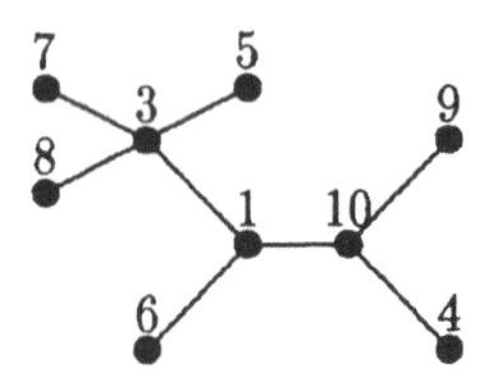

We continue by removing the leaves with the labels 4,5,6,7,8,3,1 ; and get the numbers 10,3,1,3,3,1,10.

Finally we get the sequence $s = (9, 10, 3, 1, 3, 3, 1, 10)$.

We can now establish the following property which will allow us to show that the coding is actually a bijection.

Property 1 *If* $s = (s_1, s_2, \ldots, s_{n-2})$ *is the sequence we obtained and if* i *is the smallest integer of* $[\![1, n]\!] - \{s_1, s_2, \ldots, s_{n-2}\}$ *then there exists an edge that connects the vertex with the label* i *to the one with label* s_1*.*

Proof. We are going to prove the following stronger property. If the Cayley arborescence T is transformed into s then the vertex carrying the label i is the smallest leaf of T (this leaf is connected to the vertex s_1).

We suppose first that the vertex v is an inner vertex. We denote by v_1 and v_2 two of the vertices that are connected to v by an edge. Since at the end of the transformation we come to a Cayley arborescence of two vertices, one of the vertices v_1 and v_2 has to be removed as a leaf of v during the coding. Therefore the label i has to be contained in the sequence $(s_1, s_2, \ldots, s_{n-2})$. This leads to a contradiction and therefore the vertex v has to be a leaf.

We suppose now that w is the leaf with the smallest label of T. At the beginning of the transformation, we remove this leaf. Then there remains no vertex which is connected to w, and the label of w does not show up in s. Therefore this label is greater than or equal to i. It is equal to i, since we know one leaf that carries this label: the leaf v.

We have thus demonstrated that if T is transformed into a sequence of $n - 2$ numbers between 1 and n, then the leaf removed first is v. Therefore the first element s_1 of s corresponds to the label of the vertex to which v was attached. □

In the next paragraph we will see how we can use this result to transform a sequence of $n-2$ numbers chosen from the interval $[1,n]$ into a Cayley arborescence of size n.

3.3.3 The inverse transformation

We will construct now the inverse of the transformation that we studied in the previous paragraph. Let $s = (s_1, \ldots, s_{n-2})$ be a sequence consisting of integers from the interval $[1,n]$. We want to find the Cayley arborescence that is transformed to this sequence.

For this we have to find in each step the leaf that was removed in the transformation. In the end we will be left with two vertices that were not used as leaves. We connect them to each other.

To find the ith leaf that was removed, it is sufficient to use property 1. This property says this leaf is labelled by the smallest integer of $[1,n] - \{s_i, \ldots, s_{n-2}\}$ which has not yet been used as a leaf.

Example :

Let us consider the sequence $(8,1,3,4,7,6,1,2)$. The smallest element of $[1,10] - \{8,1,3,4,7,6,1,2\}$ is 5, we connect therefore the vertex 5 to the vertex 8 :

5 8

and are left with the sequence $(1,3,4,7,6,1,2)$, and the list (5) is the list of leaves already used.

In the second step the smallest element is 8, we obtain

5 8 1

and are left with the sequence $(3,4,7,6,1,2)$ and the list $(5,8)$ of leaves already used.

In eight steps we find

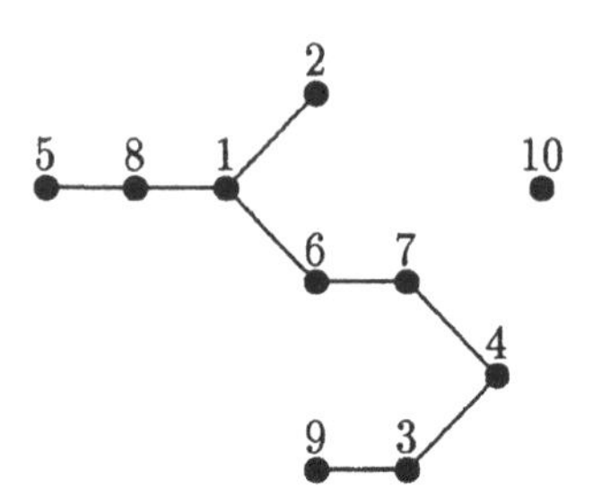

and are left with the empty sequence (), the list of leaves already used now is $(1, 3, 4, 5, 6, 7, 8, 9)$. Finally we have to connect the vertices with the labels 2 and 10, this gives

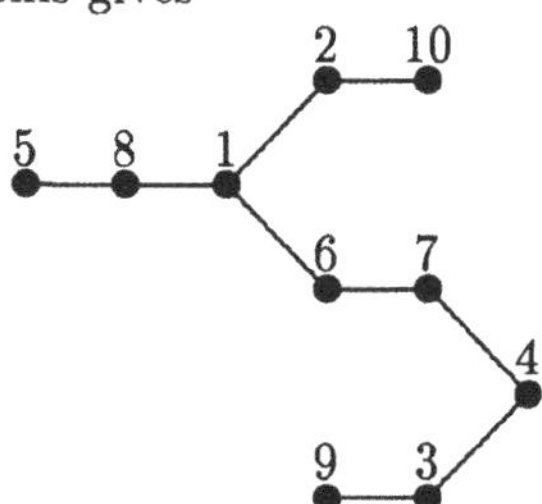

3.3.4 The algorithm

It is now easy to randomly generate a Cayley arborescence of size n. First we generate a random sequence $(s_1, \ldots, s_{n-2})$ consisting of integers from $[\![1, n]\!]$, and then we transform it into the corresponding Cayley arborescence. This transformation can be done with linear complexity using the algorithm which will be presented below. This algorithm is based on the following principle: we begin by creating a table whose i^{th} entry is equal to the number of occurrences of the integer i in $(s_1, \ldots, s_{n-2})$, and we update this table during the decoding. We also have two other variables. The first variable *maxleaf* contains the number of the largest leaf already unfolded and the second variable *newleaf* allows us to know if the next leaf will have a number smaller than *maxleaf*. These two variables are used by the function *get_next* to determine the number of the next leaf.

```
get_next ()
{
        if newleaf != 0
                return newleaf
        else
```

```
            while v[maxleaf] != 0
                  maxleaf = maxleaf+1
            maxleaf = maxleaf+1 // new position
            return maxleaf-1
}

// generate the table (s_1, ..., s_{n-2})
for i running from 1 to n - 2
    s[i] = 1+random (n) // random number between 1 and n

// calculate the table v
for i running from 1 to n
    v[i] = 0
for i running from 1 to n - 2
    v[s[i]] = v[s[i]]+1

// initial values of the variables maxleaf and newleaf
newleaf = 0; maxleaf = 1;

// construction of the Cayley arborescence
for i running from 1 to n - 2
    connect (get_next (), s [i])
    v [s [i]] = v [s [i]] - 1
    if v [s [i]] = 0 and s[i] < maxleaf
         newleaf = s [i]
    else
         newleaf = 0
// the last line
first = get_next ()
newleaf = 0
second = get_next ()
connect (first, second)
```

The calculation of the complexity of this algorithm is easy. The main part consists of four loops of complexity $O(n)$ and the function *get_next* is called n times. It only remains to observe that the loop in the function *get_next* lets the variable *maxleaf* run through the interval $[1, n]$. We have demonstrated that the above algorithm is of linear complexity.

3.4 CONCLUSION

In this chapter we have presented methods to find simple and elegant algorithms for generating certain kinds of trees in linear time. However, these will not be useful for generating more complex tree structures efficiently, therefore we will have to look for other techniques. This is the objective of the following chapters.

4

GENERATION USING BIJECTIVE METHODS

Abstract

In this chapter we show how to code a tree in some simple sequences and how to use this coding to generate easily a ternary tree with n nodes or a tree with n nodes.

4.1 INTRODUCTION

In this chapter, we present a 1-1 correspondence between the trees with n nodes and some simple words composed by letters x and y. This correspondence defined by Schützenberger [46] is constructed by going through the nodes of a tree using a postfix traversal and by replacing all nodes with k children ($k \in \mathbf{N}$) by a letter x followed by k letters y. The words that we obtain are 1-dominated words with n letters x and $n-1$ letters y.

In the first section, we explain how this bijection works. Then, in the second section, we explain in detail how this 1-1 correspondence can be used in order to obtain a random tree with n nodes and a random ternary tree with n nodes.

In the next chapter we will explain how these two algorithms can be generalized and we will propose an algorithm that generates a forest of trees split into patterns (many kinds of common trees can be generated by using this algorithm).

4.2 CODING A TREE BY A WORD

There exist many 1-1 correspondences between trees with n nodes and Dyck words with $n-1$ letters x and $n-1$ letters y. Here, we present only one of these 1-1 correspondences.This seems to be more natural in the context of generating random trees.

4.2.1 Mapping a tree T to a word $f(T)$

Let T be a tree with n nodes :

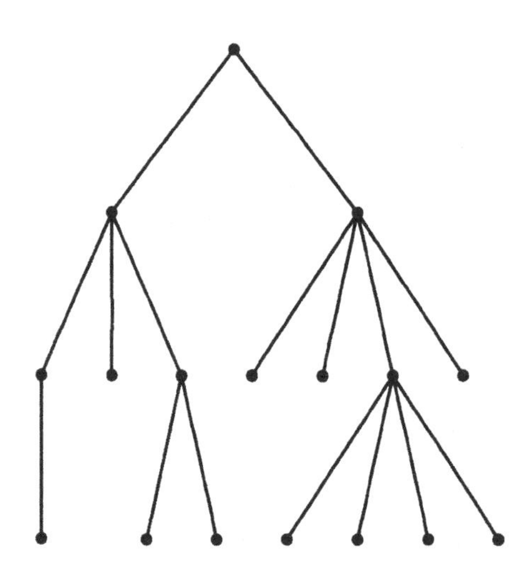

Now, we go through the nodes of T using the postfix traversal and we number the nodes accordingly. We obtain :

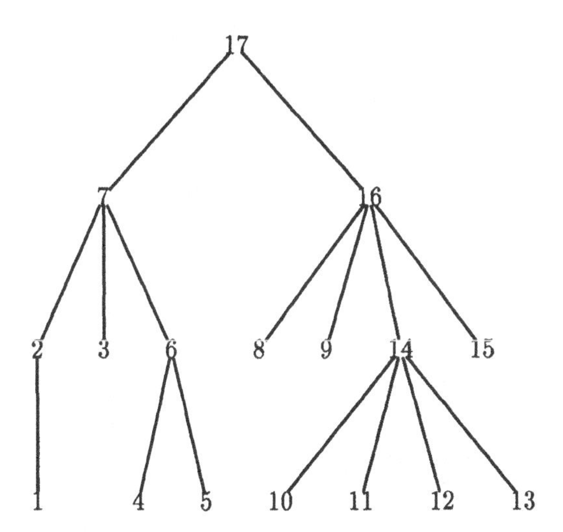

Then we take the sequence $1.2.3.\ldots.n$ and we replace in it each number i by a letter x followed by as many letters y as the node i has children. For instance the node 1 has no child so 1 will be replaced by x, the node 2 has one child so 2 will be replaced by xy, ..., the node 17 has two children so 17 will be replaced by xyy.

Therefore, we obtain a word

$$xxyxxxxyyxyyyxxxxxxyyyyxxyyyyxyy$$

that we denote by $f(T)$.

4.2.2 Properties of $f(T)$

We prove in this section that a word $f(T)$ satisfies many interesting properties. In fact, we prove that if T is a tree with n nodes then $f(T)$ is a letter x followed by a Dyck word that has $n-1$ letters x and $n-1$ letters y. We will prove in the following section that the trees with n nodes are in 1-1 correspondence with the 1-dominated words that have n letters x and $n-1$ letters y.

First, we prove a really straightforward proposition.

Proposition 2 *If T is a tree with n nodes then $f(T)$ is a word that has n letters x.*

Proof. We have replaced in the sequence $1.2.\ldots.n$, each number i by a letter x followed by some letters y. Therefore, we find in $f(T)$ as many letters x as there are numbers in the sequence $1.2.\ldots.n$ (ie. n). □

Furthermore we have the following proposition which fixes the number of letters y of $f(T)$.

Proposition 3 *If T is a tree with n nodes then $f(T)$ is a word that has $n-1$ letters y.*

Proof. The first way to proceed is to prove that there are in $f(T)$ as many letters y as there are edges in T. Indeed each time we have a node in T with k edges, we replace it by a letter x followed by k letters y in the sequence $1.2.\ldots.n$. Then we use the fact that a tree with n nodes has $n-1$ edges.

If we do not know that a tree with n nodes has $n-1$ edges, we can also proceed by induction. The proposition is true when T is a tree with one node because in this case we have $f(T) = x$.

Now we assume that the proposition is true for all trees T that have less than n nodes. Then we take a tree T with n nodes and denote by k the number of children of the root of T and by $T_1, \ldots, T_k$ the trees defined by :

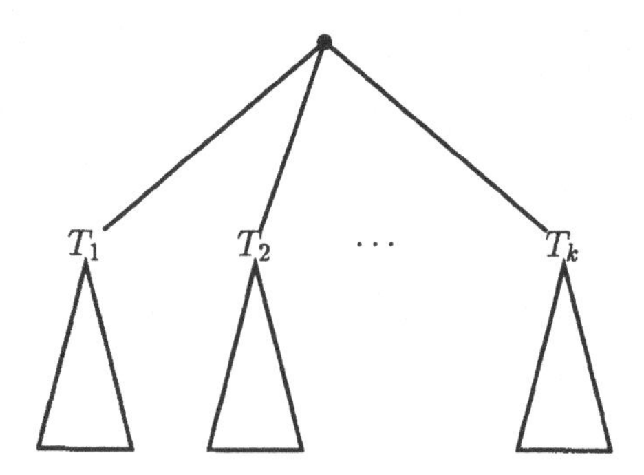

(T_i is the subtree of T whose root is the ith child of the root of T).

When we number the nodes of T, we begin by numbering the nodes of T_1 then the nodes of T_2, ..., and finally the node of T_k followed by the root of T.

Therefore, we find

$$f(T) = f(T_1).f(T_2).\ldots.f(T_k).xy\cdots y$$

where $xy\cdots y$ is a word with k letters y.

Thus by induction, we have

$$\begin{aligned} |f(T)|_y &= |f(T_1)|_y + \cdots + |f(T_k)|_y + k \\ &= (\#T_1 - 1) + (\#T_2 - 1) + \cdots + (\#T_k - 1) + k \\ &= \#T_1 + \cdots + \#T_k = \#T - 1 = n - 1. \end{aligned}$$

□

Now, we can prove the following theorem

Theorem 3 *If T is a tree of size n then $f(T)$ is a 1-dominated word with n letters x and $n-1$ letters y.*

Proof. We know already that for each tree T we have $|f(T)|_x = |f(T)|_y + 1$. We prove that $f(T)$ is a 1-dominated word by induction on the number of nodes of T.

If T has one node then we have $f(T) = x$. This *is* a 1-dominated word; so the property holds for each tree that has one node.

Now suppose that this property is true for each tree that has less than n nodes and denote by T a tree with n nodes. As in the preceding proof, we denote by k the number of children of the root of T, and by T_1, T_2, ..., T_k the subtrees of T whose root is the first, second, ..., k^{th} child of the root of T as follows:

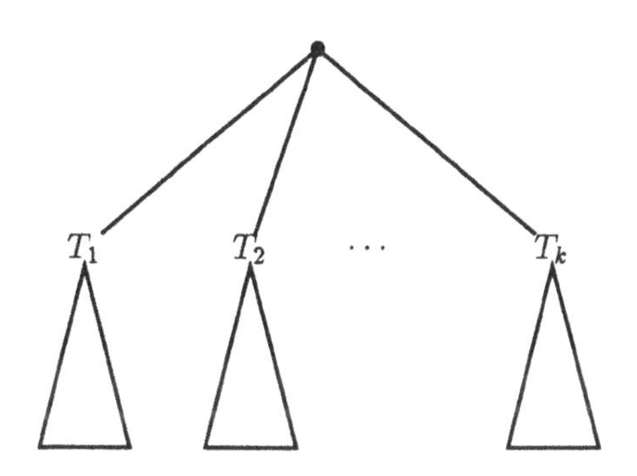

and we have

$$f(T) = f(T_1).f(T_2).\cdots.f(T_k).xy\cdots y$$

where $xy\cdots y$ is a word with k letters y.

Let v be the word $f(T)$ and v_i the prefix of v formed by the i first letters of v. We want to show that v is a 1-dominated word (ie. that for each $i \in [\![1, |v|]\!]$ we have $|v_i|_x > |v_i|_y$).

For this purpose, we choose a number i in $[\![1, |v|]\!]$. Then two different cases can arise :

- If there exists a natural integer j such that $|f(T_1)| + \cdots + |f(T_{j-1})| < i \leq |f(T_1)| + \cdots + |f(T_{j-1})| + |f(T_j)|$, we denote by l the number $i - |f(T_1)| - \cdots - |f(T_{j-1})|$ and by w the left factor of $f(T_j)$ formed by its first l letters. Then we have :

$$v_i = f(T_1).\cdots.f(T_{j-1}).w$$

and therefore

$$|v_i|_x - |v_i|_y = (|f(T_1)|_x - |f(T_1)|_y) + \cdots + (|f(T_{j-1})|_x - |f(T_{j-1})|_y) + |w|_x - |w|_y.$$

Using the induction hypothesis on the trees $T_1, T_2, \cdots, T_j$, we have $|f(T_1)|_x - |f(T_1)|_y = \cdots = |f(T_{j-1})|_x - |f(T_{j-1})|_y = 1$ and $|w|_x - |w|_y \geq 1$. Therefore, we get

$$|v_i|_x - |v_i|_y \geq j - 1 + 1 \geq 1.$$

- If i is such that $|f(T_1)| + \cdots + |f(T_k)| < i$ then we denote by l the number $i - |f(T_1)| - \cdots - |f(T_k)|$.
Then we have

$$v_i = f(T_1).\cdots.f(T_k).xy\cdots y$$

where $xy\cdots y$ is a word with $l - 1$ letters y; and

$$|v_i|_x - |v_i|_y = (|f(T_1)|_x - |f(T_1)|_y) + \cdots + (|f(T_k)|_x - |f(T_k)|_y) + (1 - (l - 1)).$$

Using the induction hypothesis on the trees $T_1, \ldots, T_k$ we get that

$$|v_i|_x - |v_i|_y = k + 1 - (l - 1)$$

is greater than or equal to 1 because $l \leq k + 1$. Indeed $k + 1$ is the number of letters that we find in the suffix $xy\cdots y$ of $f(T)$, and l is the length of a left factor of this word, so we *have* $l \leq k + 1$.

□

This theorem is crucial. It shows that if T is a tree with n nodes then $f(T)$ is a letter x followed by a Dyck word with $2n - 2$ letters. In the next section we show that indeed f defines a 1-1 correspondence between the trees with n nodes and the words formed by the concatenation of a letter x and a Dyck word with $2n - 2$ letters.

4.2.3 Decoding a 1-dominated word that has n letters x and $n - 1$ letters y

Consider a 1-dominated word v that has n letters x and $n - 1$ letters y. Here, we propose a mapping g that maps such a word v to a tree T with n nodes. Finally, we prove that g is the reciprocal mapping of f.

Now we define g by induction. If $v = x$, then we define $g(v)$ as the tree whose root has no child.

•

Suppose that we know how to compute the tree $g(v)$ for each 1-dominated word v such that $|v|_x = |v|_y + 1 \leq n$ and take a 1-dominated word v that has $n + 1$ letters x and n letters y.

Then, we map v to a path by replacing each letter x in v by a North-East step and each letter y by a South-East step. For instance if we take $v = xxyxxxxyyxyyy$, we obtain a path that begins by a North-East step because the first letter of v is x, followed by a North-East step since the second letter of v is x, ... This gives us the path :

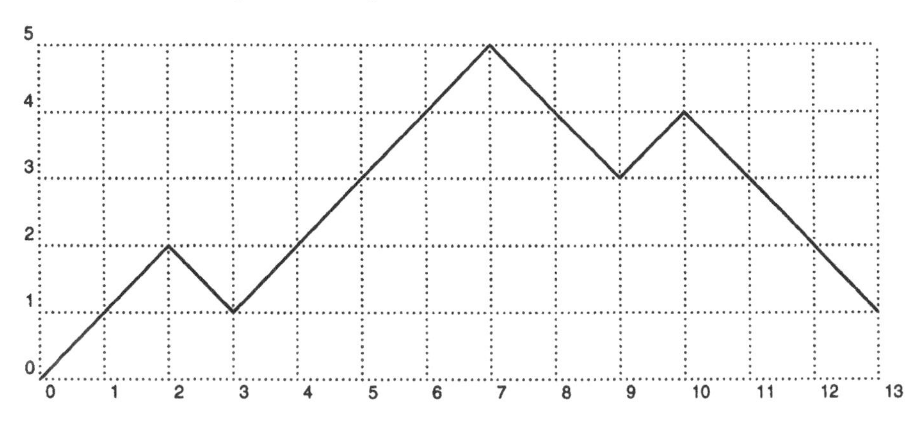

Now we denote by k the height at which the last North-East step of the path begins and by $x_1, x_2, \ldots, x_{k+1}$ the last North-East steps of the path that end at the height $1, 2, \ldots, k+1$. For instance, we have in the preceding example $k = 3$ and the steps x_1, x_2, x_3 and x_4 are defined by

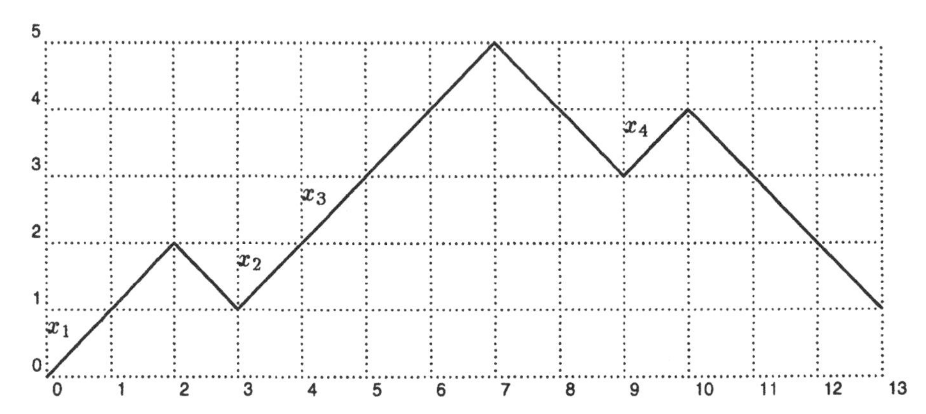

Then we partition the word v into $k+1$ words : the word v_1 formed by the letters that are between the letters x_1 (inclusive) and x_2 (exclusive), ..., the word v_k formed by the letters that are between the letters x_{k-1} (inclusive) and x_k (exclusive), and the word v_{k+1} formed by the letters that follow x_{k+1} : $xy \cdots y$. Using the same example, we obtain

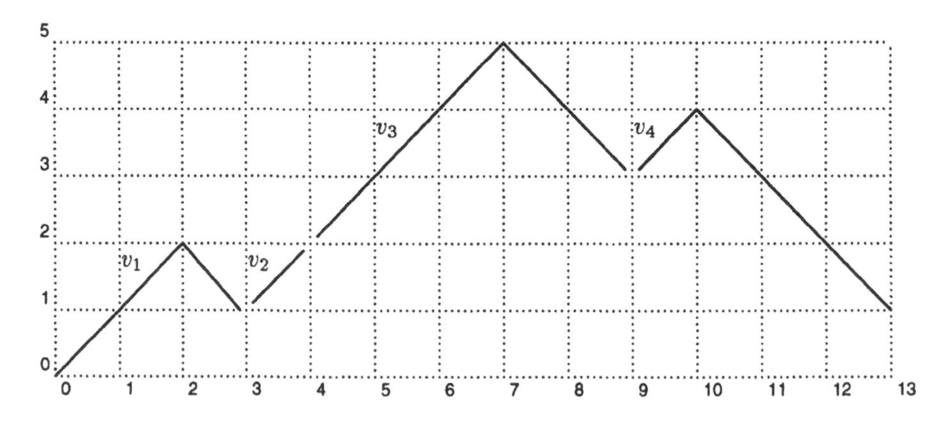

The first k words $v_1, \ldots, v_k$ are some 1-dominated words that have one letter x more than letters y. Therefore using the induction hypothesis, we find k trees $T_1, \ldots, T_k$ such that $T_1 = g(v_1), \ldots, T_k = g(v_k)$. We take for $T = g(v)$ the tree formed by a root that has the root of $T_1, \ldots$, the root of T_k for children from the left to the right.

For instance, we obtain in the preceding example

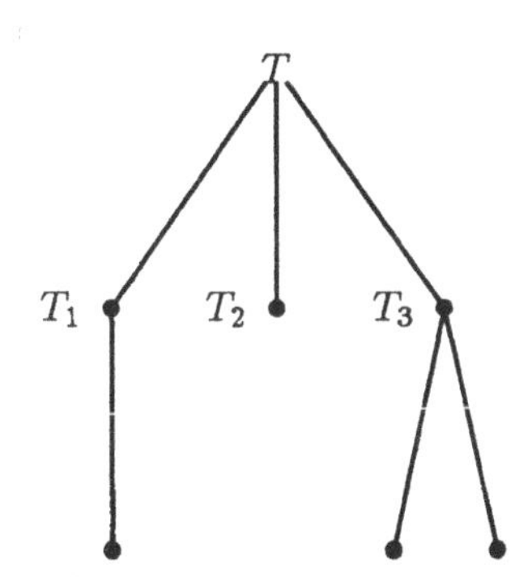

It is clear that g maps a 1-dominated word with $n+1$ letters x and n letters y to a tree T with $n+1$ nodes. Furthermore we can show the following theorem.

Theorem 4 *The mapping g defines a 1-1 correspondence between the 1-dominated words with n letters x and $n-1$ letters y and the trees with n nodes. Furthermore g is the reciprocal mapping of f.*

Proof. We need to prove that for each 1-dominated word v with n letters x and $n-1$ letters y, we have $f(g(v)) = v$ and that for each tree with n nodes we have $g(f(T)) = T$.

We proceed by induction on the number n. These two properties are true when $n = 1$. Indeed there is only one word v with one letter x and no letter y, the word x and we have $f(g(x)) = x$ and there is only one tree T with a single node and we have $g(f(T)) = T$.

Let us take an integer N and suppose that these two properties are true for each natural integer n less than N. Now take a tree T with N nodes and denote by k the number of children of the root of T and $T_1, \ldots, T_k$ the subtrees of T whose root is the first, second, ..., k^{th} child of the root of T as follows :

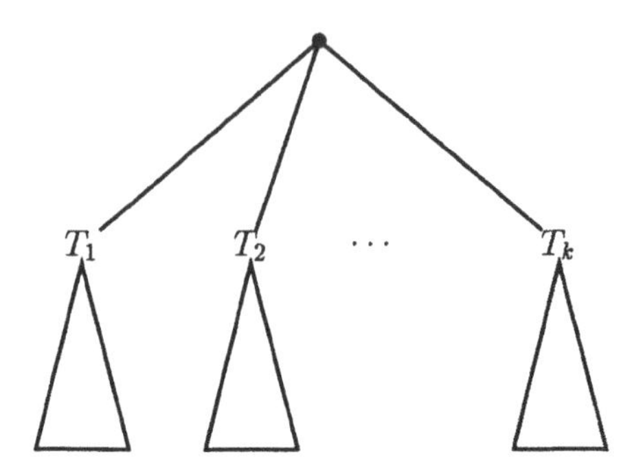

Then we have :

$$f(T) = f(T_1).f(T_2).\cdots.f(T_k).xy\cdots y$$

where $xy\cdots y$ is a word with k letters y and the words $f(T_i)$ are some words formed by a letter x followed by a Dyck word such that $g(f(T_i)) = T_i$ (by induction hypothesis).

This word can be represented by the following picture :

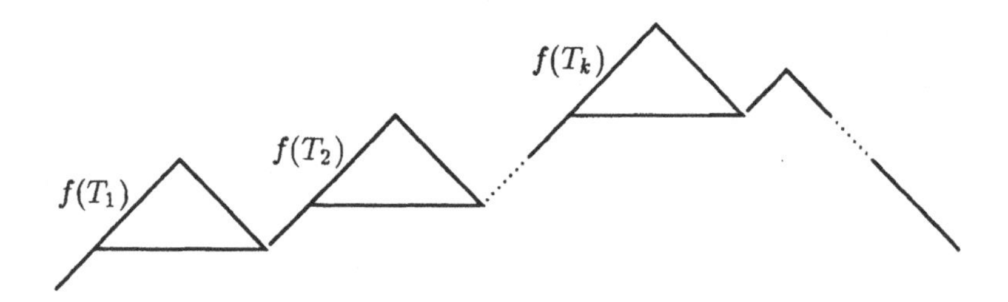

where each symbol △ represents a Dyck word.

Therefore if we denote by v the word $f(T)$ whose last letter x *begins* at height k and if we define the word $v_1, \ldots, v_{k+1}$ as in the definition of g, we find $v_i = f(T_i)$ for each i in $[\![1, k]\!]$. So g maps the word v to a tree formed by a root that has for children the roots of $T_1 = g(v_1), \ldots, T_k = g(v_k)$ from the left to the right, ie. the tree T.

Therefore, we have just to prove that if T is a tree with N nodes then $g(f(T)) = T$. The proof that v is a 1-dominated word with N letters x and $N - 1$ letters y implies $f(g(v)) = v$ is similar. □

In this section we have described the properties of a word $f(T)$ and we have presented a method to decode each 1-dominated word with n letters x and $n-1$ letters y.

4.3 TWO APPLICATIONS

We propose here two efficient algorithms that build a random tree with n nodes and a random ternary tree with n nodes. First, we see how to generate in four steps a random tree of size n. This method generates a random permutation of size $2n-1$, then a random word with n letters x and $n-1$ letters y and next a random 1-dominated word v with n letters x and $n-1$ letters y. Finally, we compute $g(v)$ in order to obtain a random tree with n nodes.

4.3.1 Generation of a permutation of size n

Let S be a set of n distinct elements $\{a_1, \ldots, a_n\}$. It is well-known that there exist $n!$ different permutations of the elements of S : $(n-1)!$ that begin with the elements a_1, ..., $(n-1)!$ that begin with the elements a_n.

Therefore, if we want to generate one of the permutations of S with uniform probability, it is sufficient to choose an element a_i of S with probability $\frac{(n-1)!}{n!} = \frac{1}{n}$ then to generate a random permutation p of $S - \{a_i\}$ and to return the permutation whose first element is a_i and the last $n-1$ ones are $p(1)$, $p(2)$, ..., $p(n-1)$.

In order to obtain a linear algorithm that generates a random permutation of the set $\{1, 2, \cdots, n\}$, we use an array S to store the elements that we can choose. At the beginning, it contains n elements, then after the choice of the first element of p, the $n-1$ first elements of S will contain the $n-1$ elements that we have not already chosen : $\{1, 2, \cdots, n\} - \{p(1)\}$, ... To maintain correctly the structure of S, we will replace at each step the element of S that we chose by the last element of S.

Therefore, a session of our algorithm can be described like this:

- At the beginning, we have $S = (1, 2, 3, 4, 5)$, $p = ()$,

- we choose an element of S, for instance the second then $p = (2)$, $S = (1, -, 3, 4, 5)$ and finally $S = (1, 5, 3, 4)$ (the last element of S goes into the free case),
- choice of the third element of S : $p = (2, 3)$, $S = (1, 5, -, 4)$ and finally $S = (1, 5, 4)$,
- choice of the third element of S : $p = (2, 3, 4)$ and $S = (1, 5)$,
- choice of the first element of S : $p = (2, 3, 4, 1)$ and $S = (5)$,
- choice of the first element of S : $p = (2, 3, 4, 1, 5)$ and $S = ()$.

This gives us the following linear algorithm to generate a random permutation of size n :

```
char S[n+1], p[n+1];
// initialization
for (i =1; i ≤ n; i++)
        S[i] = i
// random generation of p
for (i = 1; i ≤ n; i++)
        place = 1+random(n+1-i) // a random number between 1 and n+1-i
        p[i] = S[place]
        S[place] = S[n+1-i]
```

4.3.2 Generation of a random word with n letters x and $n-1$ letters y

In order to generate a random word with n letters x and $n-1$ letters y, we generate a random permutation p of size $2n-1$. Then we replace in p the numbers that are smaller than or equal to n by a letter x and the numbers that are greater than n by a letter y.

For instance, if we have $n = 3$ and $p = (2, 3, 4, 1, 5)$, we obtain the word $xxyxy$.

We have the following proposition :

Proposition 4 *If w is a word that has n letters x and $n-1$ letters y, then there exist $n!.(n-1)!$ permutations p that give the word w when we replace in p the numbers smaller than or equal to n by x and the other ones by y.*

Proof. Indeed, if we take a word w that has n letters x and $n-1$ letters y and we want to transform it into a permutation p such that the numbers smaller than or equal to n in p correspond to a letter x of w, we need first to place the numbers $\{1, \cdots, n\}$ in the n positions of w where there is a letter x. This gives us $n!$ possibilities. Then we must place the numbers $\{n+1, \cdots, 2n-1\}$ in the $n-1$ positions of w where there is a letter y, this gives us $(n-1)!$ possibilities. So the total number of permutations p is $n!$ times $(n-1)!$. □

Corollary 1 *Our algorithm draws uniformly each word with n letters x and $n-1$ letters y.*

Proof. Let us take a word w that has n letters x and $n-1$ letters y. It is drawn with a probability equal to the sum of the probabilities to draw each of the $n!.(n-1)!$ permutations p that can be mapped to w.

Therefore, it is drawn with the probability :

$$\frac{1}{(2n-1)!} + \cdots + \frac{1}{(2n-1)!} = \frac{n!.(n-1)!}{(2n-1)!}$$

since each of these permutations is drawn with a probability $\frac{1}{(2n-1)!}$. □

4.3.3 Generation of a random 1-dominated word with n letters x and $n-1$ letters y

Now we can show how to generate a random 1-dominated word with n letters x and $n-1$ letters y. First, we need to define a cyclic permutation.

Definition 12 *A cyclic permutation is a mapping which transforms each word v formed by the letters $v_1, v_2, \ldots, v_{r-1}, v_r, \ldots, v_{|v|}$ into a word $v' = v_r \cdots v_{|v|} v_1 v_2 \cdots v_{r-1}$ where r is a natural integer (i.e. the ith letter of v' is equal to the $((i+r-2) mod\,|v|+1)^{th}$ letter of v).*

Then we have the following lemma called cycle lemma [22].

Lemma 1 *There exists one cyclic permutation which transforms a sequence of n letters x and $n-1$ letters y into a 1-dominated sequence.*

Proof. Let v be a word formed by n letters x and $n-1$ letters y, and r a natural integer in $[\![1, 2n-1]\!]$. Now denote by v' the prefix of v that has $r-1$ letters and by v'' the suffix of v with $2n-r$ letters. Then we have $v = v'v''$ and if p is the cyclic permutation defined with the integer r, we have $p(v) = v''v'$.

Now we denote by X the last letter x of v that starts at a minimal height and by v_1 (respectively v_2) the prefix (respectively the suffix) of v formed by the letters that are before (respectively after) X.

For instance, if $v = xxyyyxyxxyx$, we have :

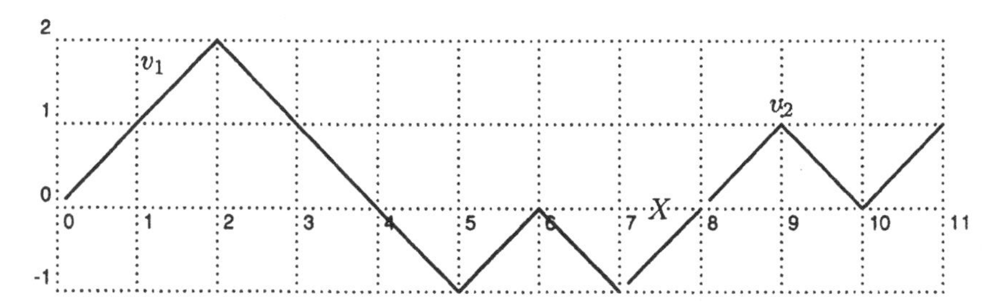

We have of course :

- $v = v_1 X v_2$,
- for each suffix v_1'' of v_1 : $|v_1''|_x \le |v_1''|_y$,
- for each prefix v_2' of v_2 : $|v_2'|_x \ge |v_2'|_y$.

Then three cases can arise :

- $|v_1| \ge r$. Then we denote by v_1' the prefix of v_1 that has $r-1$ letters and by v_1'' the suffix of v_1 that has $|v_1| - r + 1$ letters. We have $v = v_1' v_1'' X v_2$ and $p(v) = v_1'' X v_2 v_1'$.

  But
  $$|v_1''|_x - |v_1''|_y \le 0.$$
  This implies that $p(v)$ is not a 1-dominated word since $|v_1''| = |v_1| - r + 1 \ge 1$.

- $|v_1| = r - 1$. In this case, we have $p(v) = Xv_2v_1$. But this word is a 1-dominated word. Indeed let us take a prefix w of $p(v)$, then three cases can arise
  - $w = X$. In this case $|w|_x - |w|_y = 1 - 0 \geq 1$.
  - $w = Xv_2'$ where v_2' is a prefix of v_2. In this case, $|w|_x - |w|_y = |v_2'|_x - |v_2'|_y + 1 \geq 0 + 1$.
  - $w = Xv_2v_1'$ where v_1' is a prefix of v_1. Then denote by v_1'' the suffix of v_1 such that $v_1 = v_1'v_1''$, we have

$$\begin{aligned}|w|_x - |w|_y &= |wv_1''|_x - |v_1''|_x - |wv_1''|_y + |v_1''|_y \\ &= |p(v)|_x - |p(v)|_y - (|v_1''|_x - |v_1''|_y) \geq n - (n-1) + 0.\end{aligned}$$

- $|v_1| < r - 1$. Then we denote by v_2' the prefix of v_2 that has $r - 1 - |v_1|$ letters and by v_2'' the suffix of v_2 such that $v_2 = v_2'v_2''$.
  Therefore $v = v_1Xv_2'v_2''$ and $p(v) = v_2''v_1Xv_2'$. But

$$|v_2''v_1|_x - |v_2''v_1|_y = (n - |v_2'|_x - |X|_x) - ((n-1) + |v_2'|_y) \leq 0.$$

  This implies that $p(v)$ is not a 1-dominated word because $v_2''v_1$ can not be the empty word (v_2'' has at least one letter).

Therefore only the cyclic permutation that maps v to the word $w = Xv_2v_1$ gives a 1-dominated word.

We obtain in our example a word w defined by

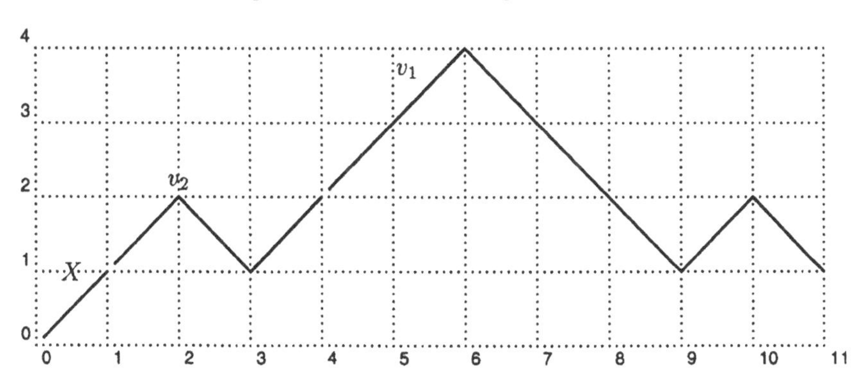

□

Now we can prove the following theorem.

Theorem 5 *There exists a 1-1 correspondence between the pairs (i, w) formed by an integer i in $[\![1, 2n-1]\!]$ and a 1-dominated word w that has n letters x and $n-1$ letters y and the words v that have n letters x and $n-1$ letters y.*

Proof. Indeed, when we have a pair (i, w) we can map it to a word v that has n letters x and $n-1$ letters y by applying to it the cyclic permutation that reads the jth letter of w to define the $(1+(i+j-2) mod(2n-1))^{th}$ letter of v. Then we obtain again the pair (i, w) in noting that the only cyclic permutation that maps v to a 1-dominated word uses the $(1+(i+j-2) mod(2n-1))^{th}$ letter of v to define the jth letter of the 1-dominated word that is w; this defines uniquely the integer i in $[\![1, 2n-1]\!]$ and the word w. □

Corollary 2 *If we want to generate a random 1-dominated word w that has n letters x and $n-1$ letters y, it is sufficient to generate a random word v that has n letters x and $n-1$ letters y, then to find the cyclic permutation p that transforms it into a 1-dominated word and finally to apply p to v.*

Proof. Indeed, if we take a 1-dominated word w that has n letters x and $n-1$ letters y, this word will be generated with a probability equal to the sums of the probabilities to generate the words $v_1, \ldots, v_{2n-1}$ where $v_1, \ldots, v_{2n-1}$ are the words that are in 1-1 correspondence with the pairs $(1, w), \ldots, (2n-1, w)$.

Therefore the probability to generate w is equal to

$$\frac{n!(n-1)!}{(2n-1)!} + \cdots + \frac{n!(n-1)!}{(2n-1)!} = \frac{n!(n-1)!}{(2n-2)!}$$

a probability that does not depend on w. □

Now we can give a linear algorithm to generate a random 1-dominated word w when we know the word v as follows

// we look first for the minimal height in v
pos = 1
height = 0
min = 0
pos_min = 1

```
run through the sequence v from the left to the right
    if the letter read is x
        if height ≤ min
                pos_min = pos
                min = height
        height = height + 1
    else
        height = height-1
    pos = pos+1
// now the variable pos points to the last letter x of minimal height
// we realize the cyclic permutation
beginning = pos
posw = 1
do
    pick the pos^th symbol of v and put it in the posw^th position of w
    pos = 1 + pos mod |v|
    posw = posw + 1
while pos ≠ beginning
```

4.3.4 Generation of a tree with n nodes

We have shown in the preceding section how to generate a random 1-dominated word v that has n letters x and $n-1$ letters y. Now we can use the 1-1 correspondence that exists between these words and the trees with n nodes.

Indeed, to build a random tree of size n, we will generate a random 1-dominated v that has n letters x and $n-1$ letters y and compute the value $g(v)$.

This value can be computed as follows :

- We initialize an empty stack.
- We read the letters of $g(v)$ from the left to the right and each time we find a letter x followed by k letters y, we take from the stack the last k trees that we have added, the trees $T_1, \ldots, T_k$, and we put into the stack the tree formed by a root that has for children the roots of the trees $T_1, \ldots, T_k$. The tree $g(v)$ is the unique tree that remains in the stack.

For instance, suppose that we have generated the word $xxyxxxxyyxyyy$, then the evolution of the stack will be

	trees in the stack
at the beginning	none
after we have read a letter x	
after we have read the letters xy	
after we have read a letter x	
after we have read a letter x	
after we have read a letter x	
after we have read the letters xyy	
after we have read the letters $xyyy$	

Now we must prove that for a given word v this algorithm generates the tree $g(v)$. For this, we use the following theorem.

Theorem 6 *Suppose that this algorithm has just read a prefix w of v. Then denote by k the height of the last letter of w, by $x_1, \ldots, x_k$ the last letters of w that end at height $1, 2, \ldots, k$ and by $w_1, \ldots, w_k$ the words formed by the letters that are before x_1 (inclusive), between x_1 (exclusive) and x_2 (inclusive), ..., between the letters x_{k-1} (exclusive) and x_k (inclusive).*

Then we find in the stack the trees $g(w_1), g(w_2), \ldots, g(w_k)$ from the bottom to the top.

Proof. We proceed by induction on the number n of letters x of the prefix w. If this number is 0 then we have $w = \epsilon$ and the stack is empty. So the theorem does hold when $n = 0$.

Now suppose that the theorem holds for a prefix of v with n letters x. Then define the words $w_1, \ldots, w_k$ as in the theorem. By induction hypothesis, we find in the stack of the algorithm the trees $g(w_1), g(w_2), \ldots, g(w_k)$ from the bottom to the top.

Now suppose that the next word that the algorithm reads is a word $xy \cdots y$ with l letters y and denote by w' the word $wxy \cdots y$. We have $l \leq k$, because w' is a 1-dominated word and we have $1 \leq |w'|_x - |w'|_y = |w|_x - |w|_y + 1 - l = k + 1 - l$.

Therefore, there are at least l trees in the stack and the algorithm can proceed. At the end of the step we find in the stack the trees $g(w_1), \ldots, g(w_{k-l})$ and the tree formed by a root that has as children from the left to the right the trees $g(w_{k-l+1}), \ldots, g(w_k)$, that is, the tree $g(w_{k-l+1} \cdots w_k xy \cdots y)$.

We conclude in noting that the height of w' is $k' = k + 1 - l$ and that when we define the word $w'_1, \ldots, w'_{k'}$ as in the theorem we find the words $w_1, \ldots, w_{k-l}$, $w_{k-l+1} \cdots w_k xy \cdots y$. □

Now, we can code this algorithm in pseudo-C.

```
create an empty stack
read the letters of v to be transformed from the left to the right
        create a new vertex w
        while the next letter of the sequence is y
              read this letter
              take the top tree from the stack and add it as the left child of w
        add the tree root at w in the stack
in the end the stack contains the tree g(v).
```

The complexity of this algorithm is of course in $O(n)$.

4.3.5 Generation of a ternary tree with n nodes

Consider a ternary tree T with n nodes. We know that this tree has only two kinds of nodes : the nodes that have three children and the nodes that have no child. Let l be the number of ternary nodes that are in T, T would then have $n - l$ leaves.

Now when we compute the word $f(T)$, we find a word that is formed by the concatenation of l words $xyyy$ and $n - l$ letters x. This word is a 1-dominated word that has n letters x and $n - 1$ letters y; therefore we must have $3l = n - 1$. Thus l is equal to $\frac{n-1}{3}$.

Now take a 1-dominated word v that is formed by the concatenation of $\frac{n-1}{3}$ words $xyyy$ and $n - \frac{n-1}{3}$ words x. We can compute the tree $g(v)$. This tree has n nodes that are of two types:

- the ternary nodes that appear when we read a word $xyyy$
- and the leaves that appear when we read a letter x that is not followed by a letter y.

Therefore, we have a 1-1 correspondence between the ternary trees with n nodes and the 1-dominated words v formed by the concatenation of $\frac{n-1}{3}$ words $xyyy$ and $n - \frac{n-1}{3}$ words x. Thus we need only draw such a word v randomly.

We proceed as for the generation of a random 1-dominated word that has n letters x and $n - 1$ letters y :

- First we generate a random permutation p of size n.
- Then we replace in p the numbers that are smaller or equal to $\frac{n-1}{3}$ by a word $xyyy$ and the remaining numbers by a letter x.

- Next, we apply to this word the cyclic permutation that transforms it into a 1-dominated word v that has n letters x and $n-1$ letters y.
- Finally, we compute $g(v)$.

All these steps have a complexity in $O(n)$, so we have obtained a linear algorithm that builds a random ternary trees with n nodes as desired.

4.4 CONCLUSION

Here we have presented the 1-1 correspondence that exists between the trees of size n and the 1-dominated words with n letters x and $n-1$ letters y.

Moreover, we have shown how to use this 1-1 correspondence to generate randomly a tree with n nodes or a ternary tree with n nodes with a uniform probability. In the following chapter, we show how to exploit the power of this 1-1 correspondence in developing a general algorithm that can build generic types of trees : the forest of trees split in patterns. This algorithm can be used to generate many kinds of forests : the forests of p trees and n nodes, the forests of p trees, n nodes and l leaves, ...

5

GENERATION OF FORESTS OF TREES SPLIT INTO PATTERNS

Abstract

We present a linear algorithm which generates randomly and with uniform probability many kinds of trees: binary trees, ternary trees, arbitrary trees, forests of p k-ary trees, ... The algorithm is based on the definition of a generic kind of tree which can be coded in the form of words. These words, in turn, can easily be generated.

5.1 INTRODUCTION

This chapter is devoted to the design of an algorithm which will can be used to generate, in linear time, more general families of trees than in the previous sections. Trees are split into some basic patterns and then coded by words which will be generated uniformly.

As in the preceding chapter we generate sequences of letters that are in 1-1 correspondence to certain classes of trees. We will code here a class of objects that were defined by Dershowitz and Zaks [21] to determine the number of certain kinds of trees. We are thus able to present an algorithm which generates the following classes of objects: forests of trees split into patterns. This class is very large and contains the trees generated in the previous chapter. It also contains arbitrary trees with n vertices and l leaves, the forests of k-ary trees, the trees with n inner vertices, an arity larger than k, and l leaves, etc..

This algorithm uses very simple principles. First, we code the class of trees that we want to generate by words in a language $\mathcal{E}$ which will be defined later. Then we generate a word of this language randomly and with uniform probability in three steps: "mixing of patterns", "introduction of missed edges", and "application of the cyclic lemma". Finally, we relate the word to the desired tree structure by the appropriate transformation.

The definition of forests of trees split into patterns is given in Section 2. Section 3 shows how to code a forest of trees by a word of $\mathcal{E}$. In Section 4 we explain how to generate such a word in linear time and how to transform it into the corresponding forest of trees split into patterns. Finally, in Section 5, we show how this algorithm can be used to generate certain classes of trees: binary trees, k-ary trees, ...

5.2 DEFINITION OF FORESTS OF TREES SPLIT INTO PATTERNS

In this section we use the notion of forests of trees split into patterns defined by N.Dershowitz et S.Zaks[20]. Each forest is defined as a combination of the structures $(F, \mathcal{M}, f)$. The first, F, is a forest of trees, this term is understood as a list of trees. The second, $\mathcal{M}$, presents a list of patterns : tree structures composed of four symbols, namely vertices and three kinds of edges. The last, f, is a mapping that sends the components of the patterns in $\mathcal{M}$ to a vertex, an edge or a set of edges of F.

First, we give the definition of the patterns, then we explain the conditions on $(F, \mathcal{M}, f)$ required for a forest of trees split into patterns.

5.2.1 Definition of patterns

We define four symbols which we will call components. These symbols are then used for the definition of a pattern.

Definition 13 *These components are*

- *the vertices represented by* $\bullet$,
- *the classical-edges represented by* $|$,

- *the semi-edges represented by* | ,
- *the multi-edges represented by* ⋮ .

We continue by defining the patterns:

Definition 14 *A pattern is a vertex or a vertex r called root that has as child a list of semi-edges, multi-edges and patterns. In the case the vertex r has a pattern M as child, there exists a classical-edge which connects r to the root of M; this classical-edge will also be considered as a child of r.*

This definition can also be given in a diagrammatical way:

Definition 15 *A pattern is*

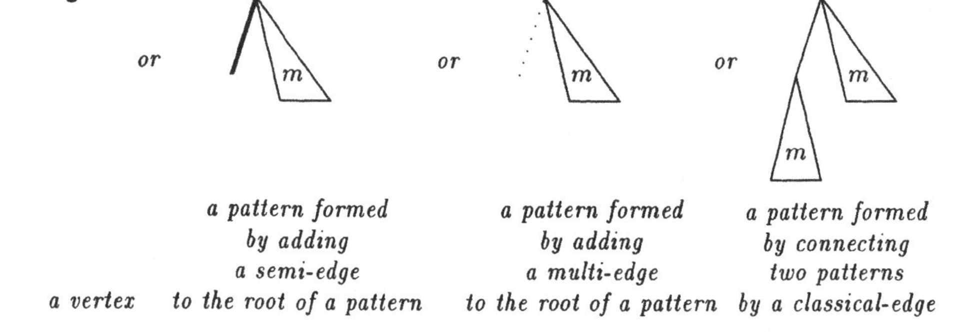

We give several examples of patterns :

Some simple patterns are :

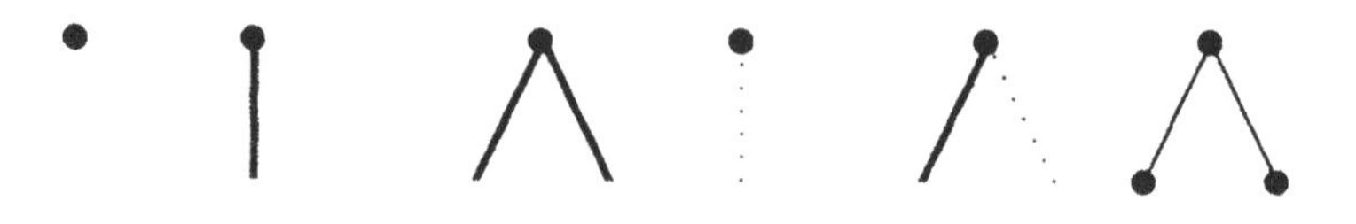

Two more complicated patterns :

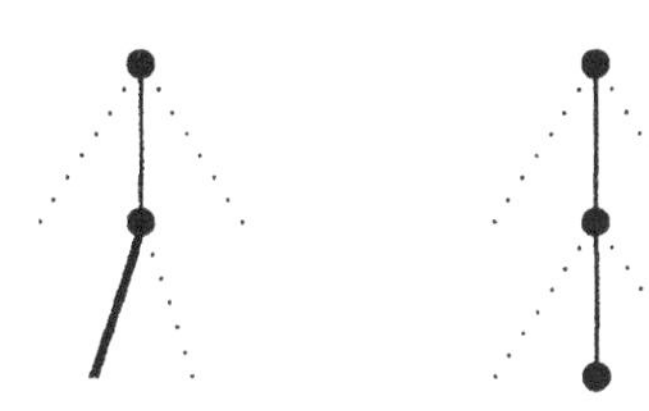

Remark :

The two first symbols (vertex and classical-edge) have almost the same name as the elements which permit the construction of normal trees : we will see, later, that these symbols will be, in a forest split into patterns $(F, \mathcal{M}, f)$, in correspondence with the vertices and edges of the forest F. The semi-edges will also be mapped to edges of F; nevertheless, they *differ* from the classical-edges by the fact that they are attached to a single vertex. A multi-edge will correspond to a set of consecutive edges of arbitrary cardinality, including zero.

5.2.2 Definition of cutting

Definition 16 *We say that the function f cuts the pattern M in a forest of trees F if and only if*

- *f associates*
  - *to each vertex of M a vertex of F,*
  - *to each classical-edge or semi-edge of M an edge of F,*
  - *to each multi-edge of M a set of edges of F that have the same parent v and that are consecutive children of v. This set can be of arbitrary cardinality, including zero.*
- *f preserves the "parent-child" relation. This means that if v and w are two items of M and if v is the parent of w in M then $f(v)$ is the parent of $f(w)$ in F.*
- *f preserves the ordering relations "left-right" order, i.e. if a vertex v of M has for edges (classical-edges, semi-edges or multi-edges) v_1, v_2, ..., v_p as children ordered from the left to the right, then the vertex $f(v)$ of F has the edges $f(v_1)$, ..., $f(v_p)$ from the left to the right, and it has* no other *edges.*

We will call (F, M, f) a forest split by a pattern if and only if f cuts the pattern M in F.

We give now some examples : Let M and $F = (T_1, T_2)$ be defined by

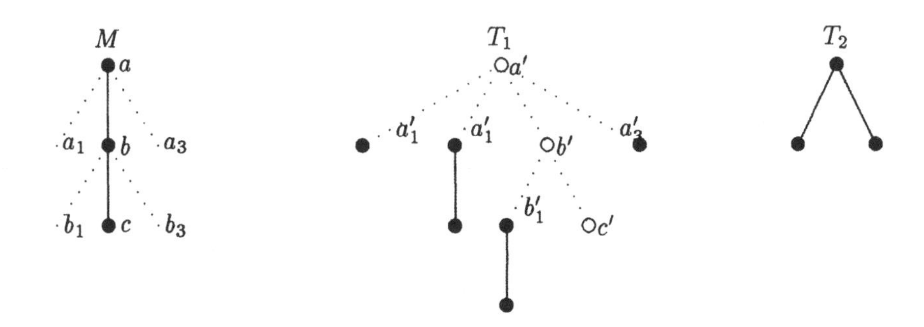

(The letters a, a_1, a_3, b, ..., c' are used to name the elements of M and F; they are not part of M or F. Certain edges and vertices of F are represented by circles or dotted lines to indicate that they belong to the set of images of the mapping f.) The mapping f that maps:

- the vertices a, b, c of M to a', b', c',
- the classical-edge that connects a to b to the edge a' b', and the edge which connects b to c to the edge $b'c'$,
- the multi-edge b_3 to the empty set,
- the multi-edges b_1 and a_3 to the sets consisting of the edge b'_1 and of the edge a'_3,
- the multi-edge a_1 to the set consisting of the two edges denoted a'_1

cuts the pattern M in the forest F.

The following is also a forest split by a pattern M when $F = (T)$ with

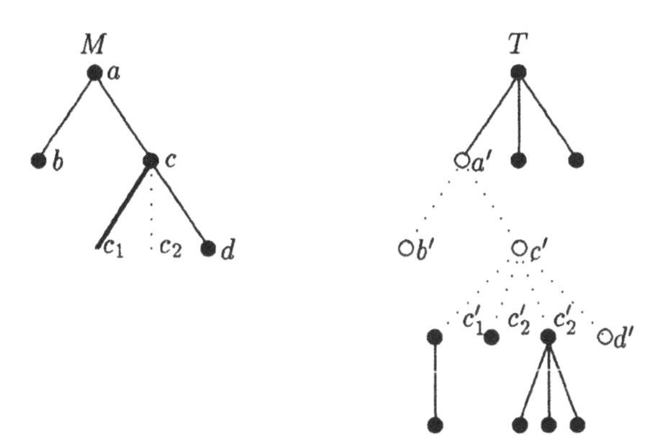

if f is the function which maps a, b, c, c_1, c_2, d to a', b', c', c'_1, c'_2, d' and the classical-edges ab, ac, cd to $a'b'$, $a'c'$ and $c'd'$.

Another way of representing a forest split by a pattern (F, M, f) is sometimes helpful. To indicate the mapping f, we mark in F the vertices and edges of F which correspond to the vertices and multi-edges of the initial pattern. The previous example can also be shown as follows :

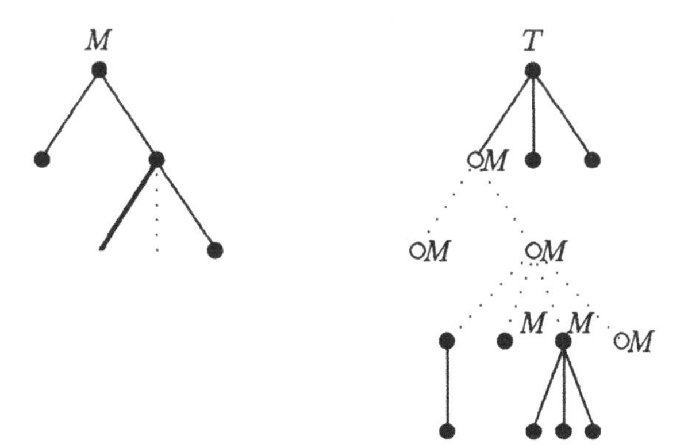

We will use this representation in the future if it defines the function f unambiguously.

5.2.3 Forests of trees split into patterns

We will now define the forests of trees split into patterns.

Definition 17 *Let M_1, ..., M_k be k patterns, a_1, a_2, ..., a_k a set of positive integers and $\mathcal{M} = ((M_1, a_1), \cdots, (M_k, a_k))$ the multiset formed by the list of patterns M_i whose multiplicity is a_i. We call $(F, \mathcal{M}, f)$ a forest of trees split into patterns if and only if*

- *$\forall i \in [\![1, k]\!]$, the restriction of f to the components of the patterns found in (M_i, a_i) cuts one by one the a_i patterns M_i in F,*
- *f is a bijective function (i.e. f maps two distinct symbols of $\mathcal{M}$ to two distinct elements (or groups of elements) of T and the image set of f is the set of elements of F).*

Example :

Consider M_1, M_2, M_3, M_4, $F = (T_1, T_2)$:

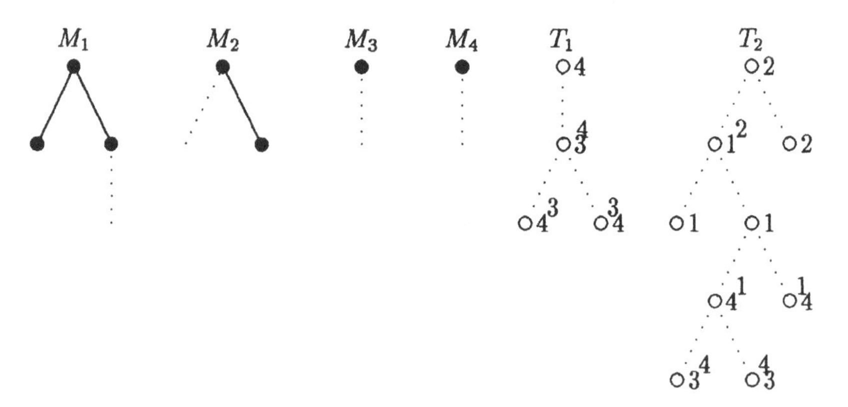

$(F, \mathcal{M}, f)$ is a forest split into patterns if we take $\mathcal{M} = ((M_1, 1), (M_2, 1), (M_3, 3), (M_4, 5))$ and denote by 1, 2, 3 and 4 the vertices and edges of F which correspond to the vertices and the multi-edges of M_1, M_2, M_3 and M_4.

The forests of trees split into patterns have the advantage that they can be enumerated easily (cf [20]). We will see that if we denote by :

- n the number of vertices found in $\mathcal{M}$,
- e the number of classical-edges found in $\mathcal{M} = ((M_1, a_1), \cdots, (M_k, a_k))$,
- c the number of semi-edges of $\mathcal{M}$,
- d the number of multi-edges of $\mathcal{M}$,
- $s = \sum_{j=1}^{k} a_j$,

the number of forests split into patterns $(F, \mathcal{M}, f)$ such that the forest F has p trees is equal to :

$$\frac{p}{s} \binom{s}{a_1, \ldots, a_k} \binom{n + d - p - e - c - 1}{d - 1}.$$

Further we will see how a forest split into patterns can be generated linearly.

Remark :

Sometimes when we get a forest F and a multiset of patterns $\mathcal{M}$ there exists a unique mapping f satisfying the conditions of Definition 17. This property can be very useful for generating the more common types of forests. Indeed suppose that we have a multiset $\mathcal{M}$ and that for all forests split into patterns $(F, \mathcal{M}, f)$ such that F has p trees, the mapping f is uniquely defined by F and $\mathcal{M}$. In this case we have a 1-1 correspondence between the forests split into patterns $(F, \mathcal{M}, f)$ and the forests F. Therefore, if we have an algorithm that builds a forest split into patterns, we will have an algorithm that builds common kinds of forests which are more interesting.

However in some cases we have several possible mappings. For instance, consider $\mathcal{M} = ((\bullet\ , 1), (M_2, 1))$ and F defined by

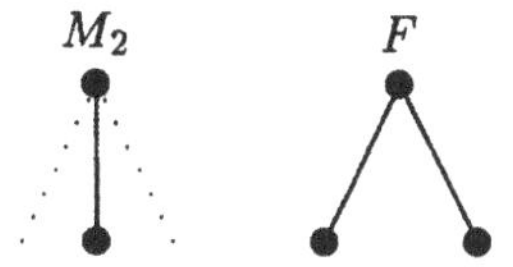

In this case, we have two possible forests split into patterns

$\circ 2$ $\quad$ $\circ 2$

or

$\circ 1^2$ $\ \circ 2$ $\quad$ $\circ 2$ $\ \circ 1^2$

5.3 CODING OF FORESTS OF TREES SPLIT INTO PATTERNS AS WORDS

First, we define two languages :

- the language $\mathcal{A}$ that contains the words formed by the symbols x, y, o, f,
- the language $\mathcal{C}$ that contains the words formed by the symbols x, y, (,), f, x_j, y^j, $)^j$, f^j, x_j^j ($j \in \mathbb{N}^\star$).

Next we have to define a language $\mathcal{B}$ that is a subset of $\mathcal{A}$. $\mathcal{B}$ will allow us to code the patterns.

There are also two subsets of $\mathcal{C}$ called $\mathcal{D}$ and $\mathcal{E}$ which are in 1-1 correspondence with some trees defined above and with forests split into patterns, respectively.

Remark :

> If we fix a multiset $\mathcal{M} = ((M_1, a_1), \ldots, (M_k, a_k))$, we will in fact work with a finite set alphabet in order to code a forest split into patterns $(F, \mathcal{M}, f)$. The words we are interested in are built from the symbols x, y, f, (,), x_1, y^1, $)^1$, f^1, x_1^1, x_2, y^2, ..., x_k, y^k, $)^k$, f^k, x_k^k.

We begin by defining a mapping that maps the patterns to a word of the language $\mathcal{B}$. Then we see that this mapping allows to define a bijection from the set of forests of trees split into patterns to a subset of $\mathcal{C}$: the language $\mathcal{E}$.

5.3.1 Coding of a pattern by a word of the language $\mathcal{B}$

A mapping t is defined recursively by applying the following rules:

- a classical-edge is mapped to y,
- a semi-edge is mapped to f,
- a multi-edge is mapped to o,

- a vertex v that has from the left to the right $(a_1, a_2, \ldots, a_p)$ as classical-edges, semi-edges or multi-edges and $(v_1, \ldots, v_{p'})$ as subtrees such that the roots of the trees $v_1, \ldots, v_{p'}$ are children of v is mapped by t to :

$$t(v_1)t(v_2)\cdots t(v_{p'})x\ t(a_p)\cdots t(a_1).$$

We obtain for example for the pattern M_1 :

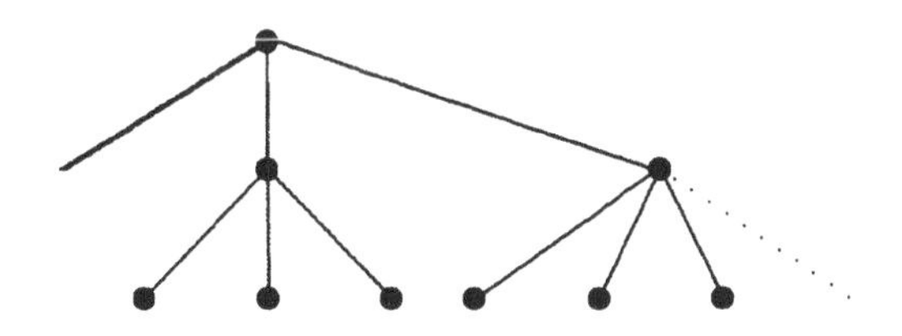

the word :

$$t(M_1) = xxxxyyyxxxxoyyyxyyf.$$

Remark :

If the pattern does not contain any multi-edge or semi-edge, we find the classical coding of a tree in the form of a sequence of the letters x, y.

Definition 18 *We denote by $\mathcal{B}$ the language defined by the grammar*

$$\mathcal{B} = x + \mathcal{B}f + \mathcal{B}o + \mathcal{B}\mathcal{B}y.$$

First we have two propositions

Proposition 5 *If $w \in \mathcal{B}$, then $|w|_x = |w|_y + 1$.*

Proof. This proposition is shown by induction using the definition of $\mathcal{B}$. □

This leads us to the following proposition :

Proposition 6 *If we remove the letters o and f from a word w of $\mathcal{B}$, we obtain a 1-dominated sequence (ie a letter x followed by a prefix of Dyck) that has one more letters x than y's.*

Proof. We proceed again by induction on the size of the word w. Using the definition of $\mathcal{B}$, we can show that, if the letters f and o are removed from a word w, we get a 1-dominated sequence. Then we use Proposition 5 to conclude. □

Using the recursive definition of the patterns (Definition 15), it is easy to see that the mapping t maps all patterns to words of $\mathcal{B}$. We give now two propositions for reconstructing the pattern M given a word $w = t(M)$ of $\mathcal{B}$.

Proposition 7 *Let M be a pattern and $w = t(M)$. If the last letter x of the word w is followed by the symbols $y_1y_2\dots y_p$, concatenation of the letters y, f, o, then*

- *the root r of M has p edges as children (classical-edges, semi-edges or multi-edges),*
- *the ith edge (classical-edge, semi-edge and multi-edge) from the right child of r is obtained by replacing the letter y_i by a classical-edge, a multi-edge or a semi-edge if y_i equals y, o or f.*

Proof. This proposition follows directly from the definition of t. □

Now we can reconstruct from a word $w = t(M)$ the edges $(a_1, \dots, a_p)$ that are children of the root of M. Next, we want to find the subtrees that correspond to the classical-edges a_i in the collection of edges $a_1, \dots, a_p$.

We define the height of a letter l in the sequence $w = t(M)$ as the difference of the number of letters x and the number of letters y that are located before l in w (the letter l inclusive).

We have the following proposition.

Proposition 8 *Let $w = t(M)$ and denote x'_i the last letter x of w that has height i. The subtree corresponding to the jth classical-edge of the root of M is coded between the letter x'_j inclusive and the letter x'_{j+1} exclusive.*

Proof. This proposition follows from Proposition 6. □

We can now prove that :

Theorem 7 *The mapping t defines a coding of patterns by words of the language $\mathcal{B}$.*

Proof. Propositions 7 and 8 allow us to construct a recursively defined mapping that maps each word w of $\mathcal{B}$ to a pattern M such that $t(M) = w$. □

5.3.2 Coding of a tree split into patterns with a word of the language $\mathcal{D}$

We define now a coding b of the trees split into patterns, then we will extend this coding to forests of trees with patterns.

The function b is actually defined on a larger set than the set of trees split into patterns :

Definition 19 *A tree cut by some patterns is a triple $(T, \mathcal{M}, f)$ for which :*

- *T is a tree,*
- *$\mathcal{M} = ((M_1, a_1), \ldots, (M_k, a_k))$ is a multiset consisting of the list of patterns M_i, each pattern appearing with multiplicity a_i. But unlike for forests split into patterns, here the a_i's are integers that can also be* equal to zero. *In that case the value of M_i is not defined.*
- *f is a bijective mapping satisfying : $\forall i \in [\![1, k]\!]$, the restriction of f to the items of patterns found in (M_i, a_i) cuts one by one the a_i patterns M_i in T.*

Remark :

> If $((T), \mathcal{M}, f)$ is a tree split into patterns then $(T, \mathcal{M}, f)$ is a tree cut by some patterns. Therefore, if we define a coding for the trees cut by some patterns, this will be a coding for the trees split into patterns.

Let $\mathcal{M} = ((M_1, a_1), \ldots, (M_k, a_k))$ be a multiset, T be a tree, and f be a function whose restriction to the components of a pattern $M = M_i$ of $\mathcal{M}$ splits this pattern in T. We define a mapping t' which depends on four arguments : a tree T, a pattern M, a mapping f and a number i.

Definition 20 *We denote t' the function where $t'(T, M, f, i)$ is a word of the language C determined by*

- *calculating $t(M)$,*
- *then replacing in $t(M)$ all the letters o that correspond to a multi-edge v of M by $card(f(v))$ letters f surrounded by the symbols (and),*
- *and finally adding to this word the index i as subscript on the first letter and as superscript on the last letter.*

Noting by "." the concatenation of two sequences, we can define the function b recursively on the trees cut by some patterns as follows :

Definition 21 *Let $(T, \mathcal{M}, f)$ be a tree cut by some patterns and denote by i the number of patterns such that f maps the roots of the patterns M_i to the root of T. Then*

- *If T is a leaf, $b((T, \mathcal{M}, f)) = t'(T, M_i, f, i)$.*
- *Otherwise, we divide the tree T in two parts : the vertices and edges of T that have as inverse image a vertex, a classical-edge, a semi-edge or a multi-edge of the same pattern M_i, and the other vertices and edges of T. This gives a set of vertices and edges T' and a set of trees that can be ordered in the same order in which their roots appear in T : $T'_1, \ldots, T'_p$ in postfix traversal. Therefore we have :*

$$b((T, \mathcal{M}, f)) = b((T'_1, \mathcal{N}_1, f_1)). \cdots .b((T'_p, \mathcal{N}_p, f_p)).t'(T, M_i, f, i)$$

where $\mathcal{N}_j$ is the multiset $((M_1, b_{1,j}), \ldots, (M_k, b_{k,j}))$ with $b_{e,j}$ the number of patterns M_e into which f splits T_j and f_j is the restriction of the function f to the components that form the patterns of $\mathcal{N}_j$.

Examples :

We start with a simple example, let M_1 be a pattern, T be a tree :

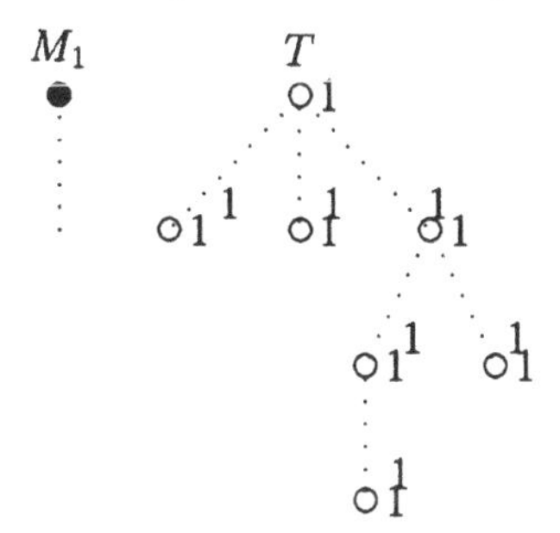

$\mathcal{M} = ((M_1, 7))$ and f be the mapping which maps seven times the root of M_1 to a node of T. Then

$$b((T, \mathcal{M}, f)) = x_1()^1 x_1()^1 x_1()^1 x_1(f)^1 x_1()^1 x_1(ff)^1 x_1(fff)^1$$

We meet here the classical representation of a tree coded by a sequence in postfix order when we remove all subscripts and superscripts, the letters (and) and when we replace the letters f by y's.

Taking the pattern M_1 and M_2 as in the preceding section, and T and f such that :

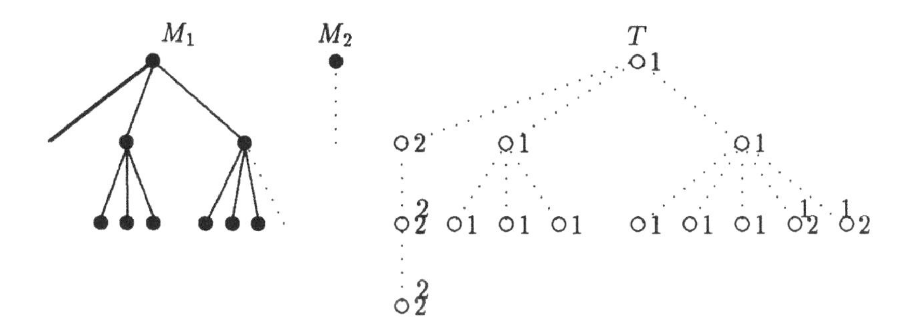

where $\mathcal{M} = ((M_1, 1), (M_2, 5))$, we get :

$$\begin{aligned} b((T, \mathcal{M}, f)) &= x_2()^2 x_2(f)^2 x_2(f)^2 x_2()^2 x_2()^2 \\ & \quad x_1 xxxyyyxxxx(ff)yyyxyyf^1 . \end{aligned}$$

With $\mathcal{M} = ((M_1, 1), (M_2, 10))$, T, and f defined by :

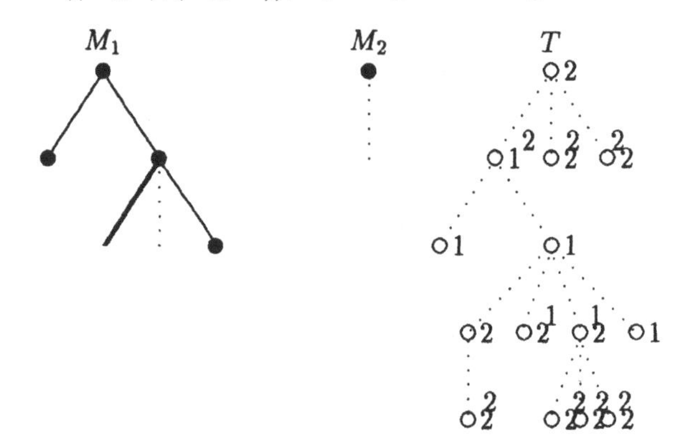

$$\begin{aligned} b((T, \mathcal{M}, f)) &= x_2()^2 x_2(f)^2 x_2()^2 x_2()^2 x_2()^2 x_2()^2 x_2(fff)^2 \\ & \quad x_1 xxy(ff)fxyy^1 x_2()^2 x_2()^2 x_2(fff)^2 . \end{aligned}$$

Finally, we look at a more complicated example with several patterns. Let us take the tree cut by some patterns $(T, \mathcal{M}, f)$ with $\mathcal{M} = ((M_1, 1), (M_2, 1), (M_3, 7))$ and M_1, M_2, M_3, T, f defined by :

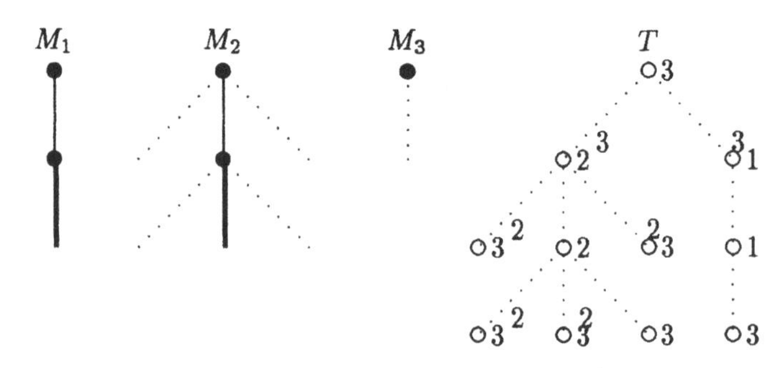

For this tree we get:

$$\begin{aligned} b((T, \mathcal{M}, f)) &= x_3()^3 x_3()^3 x_3()^3 x_3()^3 x_3()^3 x_2() f(ff) x(f) f(f)^2 \\ & \quad x x_1 f x y^1 x_3(ff)^3 . \end{aligned}$$

We can now define the language $\mathcal{D}$:

Definition 22 *The language $\mathcal{D}$ is equal to the image of the set of trees cut by some patterns under the mapping b.*

Obviously, the language $\mathcal{D}$ is a subset of the language $\mathcal{C}$. We are going to study the properties of the words of $\mathcal{D}$ now, and then we will show that b defines a $1-1$ correspondence between the trees cut by some patterns and the words of $\mathcal{D}$.

Definition 23 *We say that a word of $\mathcal{C}$ satisfies the property of the patterns if :*

- *the numbers of subscripts and of superscripts contained in the sequence are equal,*
- *the value j of the ith subscript is equal to the value of the ith superscript,*
- *the ith subscript comes before the ith superscript,*
- *the word obtained by*
  - *taking the letters between the ith subscript and the ith superscript,*
  - *removing that subscript and this superscript,*
  - *and replacing all the edges f surrounded by the symbols (and) with the letter o,*

  corresponds to a pattern M_j,
- *the patterns found are consistent i.e. two patterns with the same number are equal,*
- *there are no symbols appearing outside the representation of the patterns.*

We say that a word of $\mathcal{C}$ satisfies the property of the well-formed patterns if it satisfies the property of the patterns and if the numbers of the patterns read in this word form an increasing sequence starting at 1.

Proposition 9 *All words of $\mathcal{D}$ satisfy the property of the patterns. Moreover if $((T), \mathcal{M}, f)$ is a tree split into patterns then $b((T, \mathcal{M}, f))$ satisfies the property of the well-formed patterns.*

Proof. The proof is not difficult, this proposition is an immediate consequence of the constructive definition of the language $\mathcal{D}$. We only introduce subscripts and superscripts when we want to code a pattern; in this case, we replace in $t(M_i)$ all o's that represent a multi-edge v by a word formed by $card(f(v))$ letters f surrounded by the two letters (and). The representation of the patterns is coherent, since we always use the same multiset $\mathcal{M}$ to map the patterns into sequences.

Furthermore, if $((T), \mathcal{M}, f)$ is a tree split into patterns, we find in $b((T, \mathcal{M}, f))$ the representation of all the patterns M_i of $\mathcal{M}$. □

We define now a projection mapping Φ that maps a word of the language $\mathcal{C}$ to a sequence of the letters x and y :

- $\Phi(u.v) = \Phi(u).\Phi(v)$ if u and v are words of the language $\mathcal{C}$,
- $\Phi(u) = \epsilon$: the empty word, if u is one of the following symbols (,) or $)^j$,
- $\Phi(x) = \Phi(x_j) = \Phi(x_j^j) = x$,
- $\Phi(y) = \Phi(y^j) = \Phi(f) = \Phi(f^j) = y$.

We will say that

Definition 24 ■ *A sequence s satisfies the* dominance property *if and only if $\Phi(s)$ is a 1-dominanted sequence (i.e. $\Phi(s)$ is a word formed by a letter x followed by a left factor of a Dyck word),*

- *A sequence s satisfies the* strict dominance property *if and only if s satisfies the dominance property and if $\Phi(s)$ has one letter x more than y's (i.e. $\Phi(s)$ is formed by a letter x followed by a Dyck word)*

Proposition 10 *All the words of $\mathcal{D}$ satisfy the strict dominance property.*

Proof. We prove the Proposition 10 by induction on n, the number of vertices of T.

For $n = 1$, the proposition is true, since we must have $\mathcal{M} = ((M_1, 0), \ldots, (M_{i-1}, 0), (M_i, 1), (M_{i+1}, 0), \ldots, (M_k, 0))$ where $M_i = \bullet$ or M_i is a pattern formed by a node which has as children only multi-edges. This gives the sequences x_i^i, $x_i()^i$, $x_i()()^i$, ...

We suppose now that the proposition is true for all trees cut by some patterns $(T, \mathcal{M}, f)$ such that $_\#T < N$ and take a tree cut by some patterns $(T, \mathcal{M}, f)$ such that $_\#T = N$. We denote by i the index of the pattern that has the inverse image of the root of T as root.

Now as in the definition 21 of b, we let $T_1', \ldots, T_p'$ be the subtrees formed by the components which do not belong to the image set of M_i and T' the image under f of M_i's components and we define the multisets $\mathcal{N}_1, \ldots, \mathcal{N}_k$ and the mappings $f_1, \ldots, f_k$. It follows from the induction hypothesis that $\Phi(b((T_1', \mathcal{N}_1, f_1))), \ldots, \Phi(b((T_p', \mathcal{N}_p, f_p)))$ are p 1-dominated sequences, that have each a single letter x more than y's.

The sequence $\Phi(b((T_1', \mathcal{N}_1, f_1))). \ldots .\Phi(b((T_p', \mathcal{N}_p, f_p)))$ is therefore a 1-dominated sequence that has p letters x more than y's. We saw in the previous paragraph that the letters x's and y's that are contained in $t(M_i)$ also form a 1-dominated word. Remarking that there are exactly p letters f in $t'(T, M_i, f, i)$, we can complete the proof. □

We denote by $\mathcal{D}'$ the language that contains the words w of the language $\mathcal{C}$ which satisfy the pattern and the strict dominance properties.

We can prove the following theorem :

Theorem 8 *If w is a word of $\mathcal{D}'$, then there exists a tree cut by some patterns $(T, \mathcal{M}, f)$ such that $b((T, \mathcal{M}, f)) = w$.*

Proof. Let us take a word w of $\mathcal{D}'$. We can first extract from w the patterns M_{c_1}, M_{c_2}, ..., M_{c_k} that are encoded. This is possible in a unique way, because w satisfies the property of the patterns.

The words x_i^i, $x_i()^i$, $x_i()()^i$, ... are the only words of $\mathcal{D}'$ with one letter x. So if w has one letter x, we get $w = b((T, \mathcal{M}, f))$ using a leaf for T and $\mathcal{M} = ((M_1, 0), \ldots, (M_{i-1}, 0), (M_i, 1), (M_{i+1}, 0), \ldots, (M_k, 0))$ where M_i is a pattern formed by a root that has as many multi-edges as children as w has letters '(', and f the function which maps the root of the pattern M_i to the root of T. The lemma is therefore true for all words of $\mathcal{D}'$ with one letter x.

Assume that the theorem is satisfied for all words of $\mathcal{D}'$ that have less than N letters x and that w is a word of $\mathcal{D}'$ with N letters x. We can define a height h for each symbol r of w. This height is given by the difference of the number of letters x, x_j, x_j^j and the number of letters y, y^j, f, f^j that are located before the symbol r (inclusive). We call x_l' the last symbol x, x_j, x_j^j that has height l.

Let denote by i the superscript that is on the last letter of w. We first search for the letter x' corresponding to the beginning of this pattern. Then we split w in two parts :

- w_1 the symbols located before x' (x' exclusive),
- w_2 the symbols located after x' (x' inclusive).

Let p be the number of letters f in w_2. Then the first letter of w_2 has height $p+1$. This allows us to decompose the word w_1 in p words of $\mathcal{D}'$: $w_1 = m_1.m_2 \cdots .m_p$ where the words m_l are formed by the letters between x_l' (inclusive) and x_{l+1}' (exclusive).

We can now apply the induction hypothesis on the words $m_1, \ldots, m_p$, which are words of $\mathcal{D}'$, to obtain p trees cut by some patterns $(T'1, \mathcal{N}_1, f_1), \ldots, (T_p', \mathcal{N}_p, f_p)$ such that $b((T_1', \mathcal{N}_1, f_1)) = m_1, \ldots, b((T_p', \mathcal{N}_p, f_p)) = m_p$.

To obtain a tree cut by some patterns $(T, \mathcal{M}, f)$ that is an inverse image of w under b, we only have to map the word w_2 to a pattern M_i and then to connect the p edges that correspond to a letter f to the trees $T_1', \ldots, T_p'$. This gives a tree T. The multiset $\mathcal{M}$ is obtained by adding a pattern M_i to the multiset formed by concatenating the multisets $\mathcal{N}_1, \ldots, \mathcal{N}_p$. The function f is the function which acts on the components of $\mathcal{N}_j$ as did f_j for each j in $[\![1,p]\!]$, and maps correctly the components of M_i that were just added. □

Corollary 3 *The words of $\mathcal{D}$ are the words of C which satisfy the property of the patterns and of strict dominance.*

Proof. Propositions 9 and 10 prove that all words of $\mathcal{D}$ satisfy the strict dominance property and pattern properties.

Theorem 8 proves that for each word w of C which satisfies the patterns property and the strict dominance property there exists a tree cut by some patterns such that $b((T, \mathcal{M}, f)) = w$. To prove it, we have just defined a mapping b^{-1} which maps such a word w to a tree cut by some patterns $(T, \mathcal{M}, f)$ such that $b((T, \mathcal{M}, f)) = w$. This function is such that for each tree cut by some patterns $(T, \mathcal{M}, f)$, we get $b^{-1}(b((T, \mathcal{M}, f))) = (T, \mathcal{M}, f)$. This proves that b is bijective. □

If we restrict this function to the trees split into patterns, we get :

Corollary 4 *The mapping b defines a 1-1 correspondence between the words of $\mathcal{C}$ which satisfy the property of the well-formed patterns and of strict dominance and the trees split into patterns.*

5.3.3 Coding a forest of trees split into patterns by a word of the language $\mathcal{E}$

We will now extend the coding of trees split into patterns to forests of trees split into patterns.

Definition 25 *With the language $\mathcal{D}$ we can define the language $\mathcal{E}$ as the set of words of $\mathcal{D}^\star$ which satisfy the property of the well-formed patterns (ie. any word of $\mathcal{E}$ can be obtained by concatenating a finite number of words belonging to $\mathcal{D}$ and it satisfies the property of the well-formed patterns).*

Now we can easily code a forest of trees split into patterns $(F, \mathcal{M}, f)$ into a word of the language $\mathcal{E}$. We denote

- T_1, ..., T_p the list of trees which defines the forest F,
- $\mathcal{N}_i$ the multisets formed by the patterns of $\mathcal{M}$ that are mapped by f to the elements of T_i,
- f_i the restrictions of f to the basic items of $\mathcal{N}_i$.

Then it is sufficient, in fact, to return the word

$$b'((F, \mathcal{M}, f)) = b((T_1, \mathcal{N}_1, f_1)). \cdots .b((T_p, \mathcal{N}_p, f_p)).$$

For example, let $\mathcal{M} = ((M_1, 2), (M_2, 1), (M_3, 3), (M_4, 18))$ with :

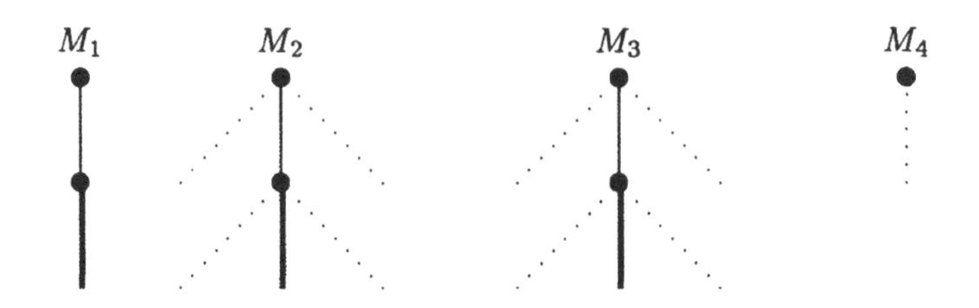

and F and f

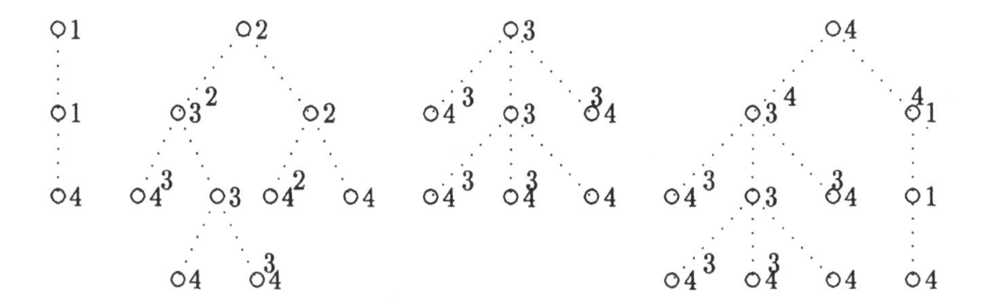

we get therefore

$$
\begin{aligned}
b'((F, \mathcal{M}, f)) = {} & x_4()^4 x_1 f x y^1 \\
& .x_4()^4 x_4()^4 x_4()^4 x_3(f) f() x() f(f)^3 x_4()^4 x_4()^4 x_2() f(f) x() f(f)^2 \\
& .x_4()^4 x_4()^4 x_4()^4 x_4()^4 x_4()^4 x_3() f(ff) x(f) f(f)^3 \\
& .x_4()^4 x_4()^4 x_4()^4 x_4()^4 x_4()^4 x_3() f(ff) x(f) f(f)^3 x x_1 f x y^1 x_4(ff)^4.
\end{aligned}
$$

Using Proposition 10 it is easy to see that we obtain a coding of forests of trees split into patterns by words of $\mathcal{E}$.

5.4 THE GENERATION ALGORITHM

We show now how a word of $\mathcal{E}$ that corresponds to a forest of p trees whose patterns are $\mathcal{M} = ((M_1, a_1), \cdots, (M_k, a_k))$ can be generated randomly and with uniform distribution.

We proceed in four steps : we begin by mixing the patterns, then we add a certain number of edges, then we use an extension of the cycle lemma [21] in order to get a word of $\mathcal{E}$, and finally we build the forest split into patterns that corresponds to this word. All these steps have a complexity in $O(n)$ where n is the number of nodes which are in $\mathcal{M}$.

5.4.1 Mixing the patterns

Let n be the number of vertices in our patterns.

We mix a_1 patterns $M_1, \ldots, a_k$ patterns M_k, this gives a number of possibilities equal to

$$\binom{s}{a_1, a_2, \ldots, a_k}$$

with $s = \sum_{j=1}^{k} a_j$. This can be easily done with the following algorithm :

```
position = 1 // this variable helps for filling the array list
for j = 1 to k
        for i = 1 to a_j
                list [position] = j
                position = position+1

// the array list contains now all non-mixed patterns

for i = position - 1 to 1 (in decreasing order)
        N = 1+random (i) // a random number of [ 1, i]
        output[i] = list[N]
        list[N] = list[i]
// the array contains the desired mixture of the patterns
```

After the mixing has been done, we replace the patterns M_i by their representations as words of C and their letters o by the word (). Thus we obtain a word of C which satisfies the property of the well-formed patterns.

The so-designed algorithm mixes uniformly the s patterns (see the Subsection 4.3.1 for a proof of the validity of this algorithm).

5.4.2 Insertion of the missed edges

In this step we add certain edges to the sequence w that we have just obtained. We introduce the following variables :

- n the number of vertices which are in $\mathcal{M}$,
- e the number of classical-edges which are in $\mathcal{M}$,
- c the number of semi-edges which are in $\mathcal{M}$,
- d the number of multi-edges which are in $\mathcal{M}$.

The forest that we wish to obtain has p trees. Therefore the corresponding word of $\mathcal{E}$ has to have $n - p$ symbols y and f. The sequence w has already $e + c$ of these symbols. We consider two cases :

If $n-p < e+c$, then we stop. There exists no forest split into patterns $(F, \mathcal{M}, f)$ such that F is a forest of p trees that has n vertices.

If $n-p \geq e+c$, then we have to add $n-p-e-c$ symbols y or f to the sequence w. But we can only add some letters f, and these letters f can be only inserted between two letters (and) which correspond to a multi-edge of a pattern, we have d possible places.

We can distribute the $n - p - e - c$ missed letters f on the d places with repetitions allowed. Therefore, we get :

$$A = \binom{s}{a_1, a_2, \ldots, a_k} \binom{n+d-p-e-c-1}{d-1}$$

different words of the language $\mathcal{C}$ (recall that $s = \sum_{j=1}^{k} a_j$).

This can be done linearly with the following algorithm :

pos = 1
number-edges $= n - p - e - c$ *// number of edges which remain to be placed*
while number_edges ≥ 1
    if random $(d - pos + number\_edges) \geq number\_edges$
        pos = pos + 1
    else

place a supplementary letter f in the pos^{th} place
number_edges = number_edges - 1

Theorem 9 *The sequences thus produced are words of C that*

- *are composed of n letters x, x_j and x_j^j, and $n-p$ letters y, y^j, f and f^j*
- *satisfy the property of the well-formed patterns,*
- *contain the patterns of the multiset $\mathcal{M}$,*
- *and begin with the representation of one of the patterns of $\mathcal{M}$.*

The algorithm generates each of the sequences that satisfy these properties with probability $\frac{1}{A}$.

Proof. We start with a sequence that has n letters x, x_j and x_j^j and $e+c$ letters y, y^j, f and f^j and add $n-p-e-c$ letters f.Therefore the resulting sequence has n letters x, x_j and x_j^j and $n-p$ letters y, y^j, f and f^j.

The sequence is formed by concatenating the s patterns of $\mathcal{M}$, then by adding f letters between the letters () in the positions that correspond to multi-edges. Therefore the sequence can only

- begin with a symbol that represents the beginning of a pattern,
- satisfy the property of the well-formed patterns,
- contain the subsequences which code the s patterns of $\mathcal{M}$.

Since we just showed that the algorithm generates A sequences with uniform probability, we only have to verify that it does not leave out any sequences. We take therefore a sequence that satisfies the properties of Theorem 9: it has subsequences that represent the a_1 patterns M_1, ..., a_k patterns M_k. Also it has n letters x, x_j and x_j^j, and satisfies the property of the well-formed patterns. This sequence can therefore be produced by first mixing the s patterns of $\mathcal{M}$, then by adding $n-p-e-c$ letters f in the d places that correspond to the d multi-edges of $\mathcal{M}$. So it can be constructed with our algorithm. □

5.4.3 Application of the cycle lemma

The sequence w that we just constructed in the previous paragraph satisfies all properties of the words of $\mathcal{E}$ except for one : the dominance property.

We will now show how a sequence with n letters x, x_j, x_j^j and $n-p$ letters y, y^j, f, f^j can be transformed into a word of $\mathcal{E}$.

For this, we recall the definition of the cyclic permutations :

Definition 26 *A cyclic permutation is a mapping which maps a word v formed by the symbols a_1, a_2, ..., a_{r-1}, a_r, ..., $a_{|v|}$ to a word $v' = a_r \cdots a_{|v|} a_1 \cdots a_{r-1}$, where r is an integer from $[\![1, |v|]\!]$ (i.e. the ith symbol of v' is equal to the $((i+r-2) \bmod |v| + 1)^{th}$ symbol of v).*

First, we prove the following lemma which is an extension of the cycle lemma 1 (see page 44).

Lemma 2 *There exist p cyclic permutations that transform a sequence of n letters x and $n-p$ letters y into a 1-dominated sequence.*

Proof. This lemma appeared first in [22], a paper by A. Dvorezky and T. Motzkin. Since then many proofs have been found (see also [21] for the history). We will give only one here: the proof of [21] which was inspired by the paper of Silberger [47].

It consists of showing that removing two consecutive letters, x followed by a y, does not change the number of cyclic permutations that we are looking for. By repeated application of this procedure, we get a sequence that consists of p letters x. This sequence can be transformed into a 1-dominated sequence (i.e. one x followed by a left Dyck factor) by p cyclic permutations. □

We have then the following theorem

Theorem 10 *There exist p cyclic permutations that transform the sequence w into a word of $\mathcal{E}$.*

Proof. Consider the given sequence w and denote $u = \Phi(w)$ (where Φ is the mapping defined in the previous paragraph that projects the letters x, x_j, x_j^j to the letter x and the letters y, y^j, f and f^j to y). Theorem 9 implies that the sequence u consists of n letters x and $n - p$ letters y. We can apply the Lemma 2 to show that there exist p cyclic permutations which transform u into a 1-dominated sequence.

We can now look for cyclic permutations that transform w into a sequence that satisfies the dominance property and begins with a letter x, x_j or x_j^j. Let t be one of these permutations. The word $\Phi(t(w))$ is then a 1-dominated sequence that can be obtained by a cyclic permutation t_1 acting on the word u. If we require also that the first letter x of $t(w)$ corresponds via Φ to the first letter of $t_1(u)$, then the cyclic permutation t_1 is uniquely determined by the choice of t.

We can also show that the choice of a cyclic permutation t_1 determines uniquely a cyclic permutation t that transforms the word w into a word $t(w)$ which satisfies the dominance property. We have therefore as many cyclic permutations that transform w into a word satisfying the dominance property as we have cyclic permutations that transform u into a 1-dominated word, i.e. p.

It only remains to be seen that these p cyclic permutations indeed give words of $\mathcal{E}$. We denote by w' a word of $\mathcal{C}$ that was obtained by applying to w one of the p cyclic permutations. The sequence w' satisfies the dominance property and begins with one of the letters x, x_j or x_j^j. We still have to check that w' satisfies the property of the well-formed patterns and that the patterns contained in w' correspond to the list $\mathcal{M}$. We use Proposition 6 (page 62) for this. This proposition shows that the representation of a pattern in the form of a word of $\mathcal{C}$ satisfies the property of strict dominance. This implies that it is impossible that the representation of a pattern is cut in two by the transformation of w into w'. □

The third step of our algorithm is to choose randomly one of the p cyclic permutations that transform w into a word of $\mathcal{E}$ (with probability $\frac{1}{p}$). Then this transformation is applied to w. This gives us :

// we look first for the minimal position of the sequence
pos = 1
height = 0
min = 0

```
pos_min = 1
run through the sequence from the left to the right
    if the letter read is x, x_j, x_j^j
        if height ≤ min
                pos_min = pos
                min = height
        height = height + 1
    if the letter read is y, y^j, f, f^j
        height = height-1
    pos = pos+1
// the variable pos now points to the last letter x, x_j, x_j^j of minimal height
// and min indicates this height
// we choose one of the p possible cyclic permutations,
height_chosen = min+random(p)
// we search for the first letter of the new sequence
pos = 1
height = 0
run through the sequence from the left to the right
    if the letter read is x, x_j, x_j^j
        if height = height_chosen
                beginning = pos
        height = height+1
    if the letter read is y, y^j, f, f^j
        height = height-1
    pos = pos+1
// we realize the chosen cyclic permutation
pos = beginning
posw = 1
do
    pick the pos^th symbol of w and put it in the posw^th position of the new sequence
    pos = 1 + pos mod |w|
    posw = posw + 1
while pos ≠ beginning
```

Theorem 11 *We obtain each word of $\mathcal{E}$ which corresponds to a forest of p trees $(F, \mathcal{M}, f)$ with probability $\frac{1}{A}$:*

$$A = \frac{p}{s} \binom{s}{a_1, a_2, \ldots, a_k} \binom{n+d-p-e-c-1}{n-p-e-c}.$$

Proof. Here it is sufficient to show that to each word of $\mathcal{E}$ corresponding to a forest of p trees $(F, \mathcal{M}, f)$ correspond s sequences which satisfy the properties of Theorem 9 (page 75) and then to apply Theorem 10.

We take a word w' of $\mathcal{E}$ that corresponds to a forest split into patterns $(F, \mathcal{M}, f)$ such that the forest F has p trees. This word has s letters x_j, x_j^j which indicate the beginning of a pattern. We denote these letters by $l_1, \ldots, l_s$. Then we define the s cyclic permutations that transform the sequence w' into a sequence that begins with one of the s letters $l_1, \ldots, l_s$. These s cyclic permutations transform w' into sequences w that satisfy the properties of Theorem 9 (page 75). They are also the only cyclic permutations that transform w' into a word whose first symbol corresponds to the beginning of a pattern. □

We have therefore the following corollary :

Corollary 5 *The number of forests of $(F, \mathcal{M}, f)$ where F is a forest of p trees, is equal to :*

$$A = \frac{p}{s} \binom{s}{a_1, a_2, \ldots, a_k} \binom{n+d-p-e-c-1}{d-1}.$$

Proof. We use the 1-1 mapping between the forests split into patterns and the words of $\mathcal{E}$ and then Theorem 11. □

5.4.4 Mapping words of $\mathcal{E}$ to forests split into patterns

In this section we present a decoding algorithm that can map a word w of $\mathcal{E}$ to a forest F that is the first ingredient of the forest split into patterns $(F, \mathcal{M}, f) = b^{-1}(w)$. Indeed, this is the unique ingredient of the forest split into patterns that we need when we want to build some common kinds of forests. This algorithm reads a word of $\mathcal{E}$ and builds the forest F. It is linear and uses a data structure which is similar to a stack in order to store the constructed trees. We use two pointers. The first, *stackptr*, points always to the top of the 'stack', and the second *patternptr* keeps the position of the 'stack' at the entrance of a pattern.

create an empty stack
read the word of $\mathcal{E}$ *to be transformed from left to right*
    if we are at the beginning of a pattern (letter x_j *or* x_j^j*)*
        patternptr = stackptr
    create a new vertex v
    while the next symbol of the sequence differs from x, x_j, x_j^j
        read this symbol
        if the symbol is equal to f *or* f^j
            delete a tree of patternptr and add it to the children of v
        if the symbol is equal to y *or* y^j
            delete a tree of stackptr and add it to the children of v
    add the created subtree in the stack
    if we are at the end of a pattern (letter x_j^j, y^j *or* f^j*)*
        stackptr = patternptr
in the end the stack contains the subtrees $T_1, \ldots, T_p$ *that form the forest* F.

5.5 APPLICATIONS

We will now show how to generate some common kinds of trees with the algorithm designed in the previous sections.

5.5.1 Generating binary trees of size $2n + 1$

Proposition 11 *There exists a 1-1 mapping between binary trees with* $2n + 1$ *vertices and the trees split into patterns* $((T), \mathcal{M}, f)$ *such that :*

- $\mathcal{M} = ((\bullet\ , n+1), (M_2, n))$ *with* M_2 *equal to*

- T *is a tree of size* $2n + 1$.

Proof. Let T be a binary tree with $2n + 1$ vertices. Then it is easy to see that T has n inner vertices and $n + 1$ leaves.

If we try to define a function f that splits the patterns of $\mathcal{M}$ in T, then we have only one choice. We have to associate $n+1$ times a leaf of T to the vertex of the pattern M_1 and n times an inner vertex of T and the edges that are its children to the components of the patterns M_2.

Example :

If we take $\mathcal{M} = ((M_1, 3), (M_2, 2))$ and T as indicated below :

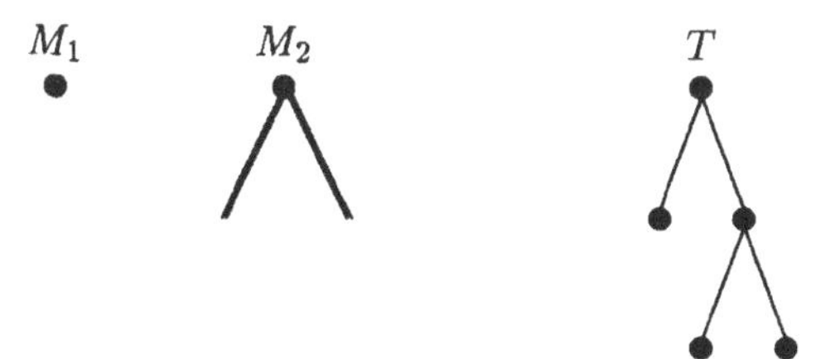

then there exists only one tree split into patterns $((T), \mathcal{M}, f)$:

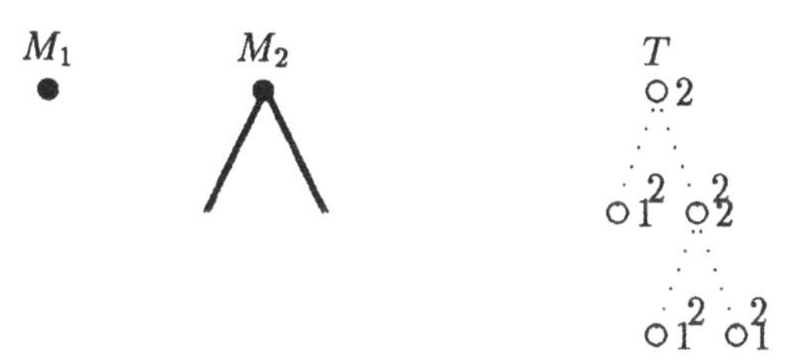

□

This theorem and the corollary 5 (page 79) recover a classical result: the number of binary trees of size $2n+1$ is equal to

$$\frac{1}{2n+1}\binom{2n+1}{n},$$

the well known Catalan numbers.

In addition we obtain a second linear algorithm that generates uniformly a binary tree of size $2n+1$. Indeed, we only have to generate a tree split into patterns $((T), \mathcal{M}, f)$ with $\mathcal{M} = ((\bullet\ , n+1), (M_2, n))$ and then keep only the tree T.

5.5.2 Generating forests with p k-ary trees and $kn+p$ vertices

We only have to exploit the fact that these forests are in 1-1 correspondence with the forests split into patterns $(F, \mathcal{M}, f)$ where F is a forest with p trees and $\mathcal{M} = ((\bullet\ , (k-1)n+p), (M_2, n))$. Here M_2 stands for a pattern corresponding to a k-ary vertex:

The number of these forests is equal to

$$\frac{p}{kn+p}\binom{kn+p}{n}.$$

5.5.3 Generating trees with n vertices

This time we use the pattern list $\mathcal{M} = ((M_1, n))$ where M_1 is the pattern representing a vertex of arbitrary arity

The number of these trees is

$$\frac{1}{n}\binom{2n-2}{n-1}.$$

The number of trees with n nodes is therefore equal to the number of binary trees with $n-1$ inner nodes. There are many 1-1 correspondences between these two families of trees (see the Subsection 3.1.3).

5.5.4 Generating forests of trees with n inner vertices and l leaves

We take $\mathcal{M} = ((M_1, n), (\bullet \;, l))$ with

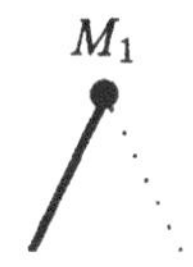

and for F a forest of p trees.

The number of these forests is

$$\frac{p}{n+l}\binom{n+l}{n}\binom{n+l-p-1}{l-p}.$$

5.5.5 Generating forests with p unary-binary-ternary trees with a leaves, b unary vertices, c binary vertices, d ternary vertices if $a = c + 2d + p$

We use, this time, $\mathcal{M} = ((\bullet \;, a), (M_2, b), (M_3, c), (M_4, d))$ with

and for F a forest of p trees.

The number of these forests is

$$\frac{p}{a+b+c+d}\binom{a+b+c+d}{a, b, c, d}. \tag{5.1}$$

5.6 CONCLUSION

We have designed a linear algorithm that randomly generates a forest of trees split into patterns. This algorithm allows us to randomly generate the trees that we encounter often in "everyday life" : k-ary trees, arbitrary trees, trees with n inner vertices and l leaves, ... This algorithm can be implemented on parallel computers, and allows us then to generate a word of $\mathcal{E}$ which corresponds to a tree split by a pattern with a complexity of the order $O(Log^2(n))$ (see Chapter 10).

But, unfortunately, it does not allow us to generate all kinds of trees linearly. For example, we will see in the next chapter how colored trees can be generated by using a 1-1 correspondence between these trees and some kind of forest. Moreover, all classes of forests that we can generate can be enumerated by simple formulas. For instance, if we want to generate a unary-binary tree, we need first to choose the number of binary vertices in the tree; a method to choose such a number uniformly is proposed in the Chapter 7. These trees are in bijection with the Motzkin words and many other combinatorial different objects. For this reason, they are very interesting. Also, the formulas for their enumeration are not as simple as those we have seen in this paper (the use of a summation $\sum$ is necessary).

6

GENERATION OF COLORED TREES

Abstract

We present a linear time algorithm which generates uniformly a colored tree. Previous methods do not apply directly for the generation of these trees. We use a one-to-one correspondence between colored trees and sequences which are then uniformly generated in linear time.

6.1 INTRODUCTION

Colored trees are a special class of trees whose random generation does not fit directly into the scope of the methods developed in the previous chapters. The organization of this chapter is as follows : in Section 2 we recall the definitions and basic properties of colored trees, in Section 3 we present a bijection between forests of k colored trees and some sequences while in Section 4 we show how to generate these sequences in linear time. Concluding remarks and further aspects are offered in Section 5.

6.2 FORESTS OF COLORED TREES

In this section we give the definition of colored trees and some basic properties.

Definition 27 ■ *A black tree is either a black leaf or a tree whose root is black and whose root has a single child that is a white tree.*

- *A white tree is either a white leaf or a tree whose root is white, the left child of the root being a white tree and the right child a black tree.*
- *A colored tree is a white tree.*

Examples :

We display below the five colored trees whose height is less than or equal to three.

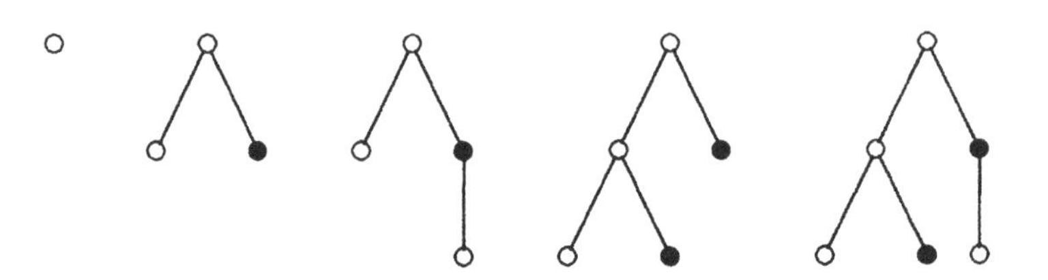

Property 2 *A forest of l colored trees with m black nodes and n white nodes has*

- $2m - n + l$ *black leaves,*
- $n - m - l$ *black unary nodes,*
- $n - m$ *white leaves,*
- m *white binary nodes.*

Proof. We proceed by induction on the size of the forests of colored trees. If $m + n = l$, the only existing forest with l colored trees is the forest with l white leaves. We get then $m = 0$, $n = l$. This forest has no unary black node, no black leaf, no binary white node. This property is therefore true for $m + n = l$.

Assume that the property is fulfilled for all forests of l trees with m black nodes and n white nodes such that $l \leq m + n < N$ and take a forest F with l colored trees such that $m + n = N$. Since $N > l$, there exists a tree T_i of the forest F with at least two nodes. This tree T_i has at least one node v which fulfills one of the two following properties :

- v is a unary black node with a white leaf as child,
- v is a binary white node with a white leaf and a black leaf as children.

Suppose that such a node v of F has been chosen,

- if v is a black node, replace in T_i the node v by a black leaf, we get a new colored tree T_i' :

  Example :

  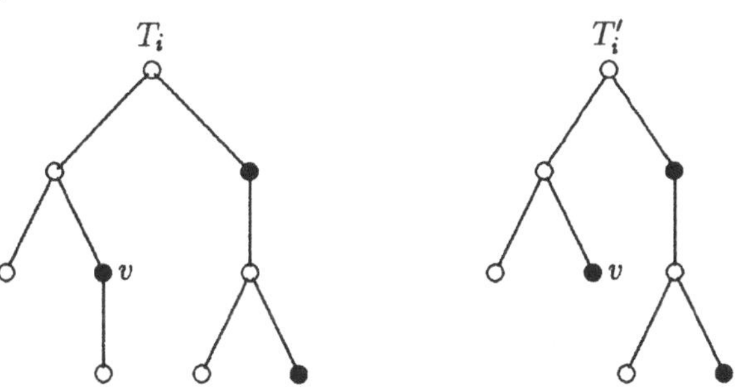

  By replacing T_i with T_i', we get a colored forest F' with l trees. F' has m black nodes and $n-1$ white nodes. Thanks to the induction hypothesis, F' has $2m-n+1+l$ black leaves, $n-m-1-l$ unary nodes, $n-m-1$ white leaves and m binary nodes. F has now one unary node and one white leaf more than F' but one black leaf less F'. This ends the proof.

- If v is a binary node, T_i' derived from T_i by replacing the subtree of v by a white leaf is a colored tree :

  Example :

  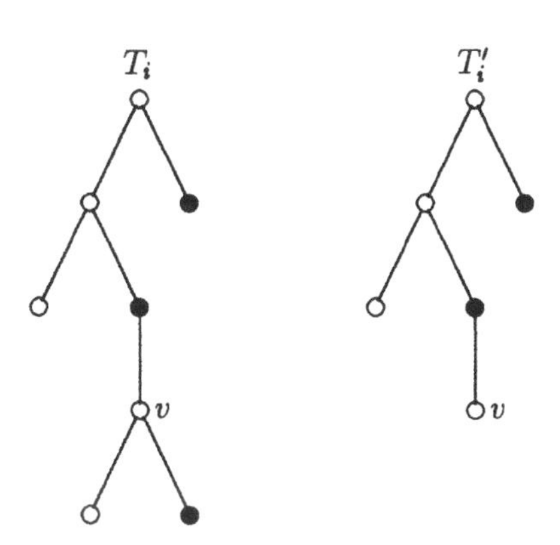

The forest F' obtained by replacing T_i by T_i' in F has l colored trees with $m-1$ black nodes and $n-1$ white nodes. Applying the induction hypothesis to F', we see that this forest has $2m-n+l-1$ black leaves, $n-m-k$ unary nodes, $n-m$ white leaves and $m-1$ binary nodes. The fact that F has one black leaf and one binary node more than F' permits to conclude.

□

6.3 BIJECTION

In this section, we define a list of black and white trees which we denote $W(p,q,k)$. Then we present a 1-1 correspondence between the colored forests with k trees having p black nodes, q white nodes in which a white node was labelled, and $W(p,q,k)\times[\![1,k]\!]$ (ie. the pairs formed with an element of $W(p,q,k)$ and a natural integer of $[\![1,k]\!]$).

6.3.1 Some definitions

We assume throughout this chapter that p, q and k are positive integers such that : $p+k \leq q \leq 2p+k$.

Definition 28 *$W(p,q,k)$ is the set of sequences composed of p white trees and $p+k$ black trees which fulfill the following properties :*

- *all white trees are leaves,*
- *$2p-q+k$ black trees are leaves and the remaining are trees with a black root whose child is a white leaf,*
- *a black leaf is always preceded by a white leaf.*

Example :

Here is a sequence of $W(3,6,2)$:

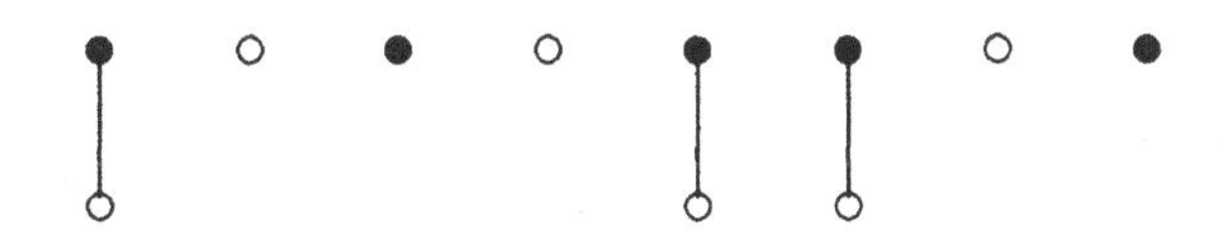

But there exists no triple (p, q, k) such that $W(p, q, k)$ contains the sequence :

because the last black leaf of this sequence is preceded by a black tree : in fact the tree .

Definition 29 *$N(m, n, l)$ is the set of colored forests with l trees with m black nodes and n white nodes in which a white node has been chosen.*

Then we call pointed forest the elements which belong to the union of the $N(m, n, l)$ where $m \in \mathbf{N}$, $l \in \mathbf{N}^\star$ and $n \in [\![m+l, 2m+l]\!]$. We will add a star $\star$ on the drawings in order to make a distinction between the chosen white node and the others.

6.3.2 A bijection between $W(p, q, k) \times [\![1, k]\!]$ and $N(p, q, k)$

We define first a mapping t which associates to each element $w = (s, n)$ in $W(p, q, k) \times [\![1, k]\!]$ a pointed forest of $N(p, q, k)$. Then we will prove that t is a bijection.

Definition of t

In this section we define five maps t_1, t_2, t_3, t_4, t_5 which send progressively an element of $W(p, q, k) \times [\![1, k]\!]$ to a pointed forest of $N(p, q, k)$. The function t will then be defined by composition of these maps.

Definition 30 *t_1 is the mapping which transforms an element $s \in W(p, q, k)$ into a sequence s' derived from s by :*

- *adding a star ($\star$) to the white leaf which belongs to the first tree of s,*
- *applying the cyclic permutation which puts the last tree of s in the last position.*

Example :

If s is defined by

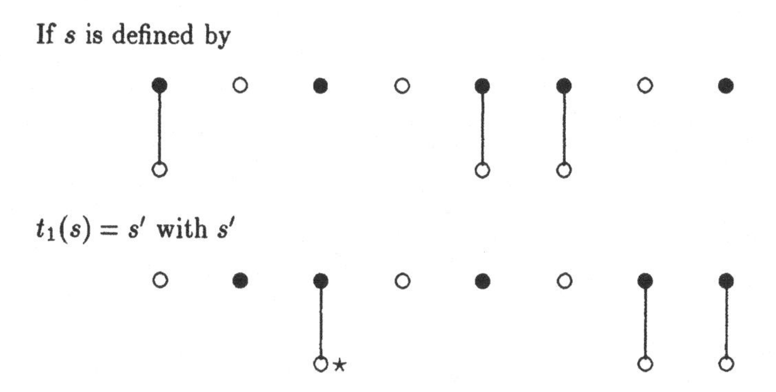

$t_1(s) = s'$ with s'

Remark :

The definition of t_1 is correct since if $p + k \leq q$, all sequences of $W(p, q, k)$ contain at least k ($k \geq 1$) trees .

Let Φ be the function which transforms a forest s' containing a white node with a star ($\star$) into a forest s' in which the starry white node has been replaced by a white node.

We get then the following result :

Theorem 12 *If $s \in W(p, q, k)$ and if $s' = t_1(s)$ then $\Phi(s')$ is a sequence of $W(p, q, k)$ ending with .*

Proof. Since the forest $\Phi(s')$ is obtained through a cyclic permutation applied to s, $\Phi(s')$ has the right number of ○ , ● , . By construction, $\Phi(s')$ ends with the tree therefore all black leaves of $\Phi(s')$ are preceded by a white leaf. □

Take now a forest $s' = t_1(s)$ and let $a_1, a_2, \ldots a_{q-p}$ be the increasing sequence of integers indicating the positions of the $q-p$ trees which compose s' and $a_0 = 0$. We can split s' in $q-p$ pieces : $s' = s'_1 s'_2 \cdots s'_{q-p}$ where s'_i is the forest of the trees that are between the a^{th}_{i-1} (exclusive) and the a^{th}_i trees of s'.

Example :

With the sequence s' of the preceding example, we get : $a_0 = 0, a_1 = 3, a_2 = 7, a_3 = 8$.

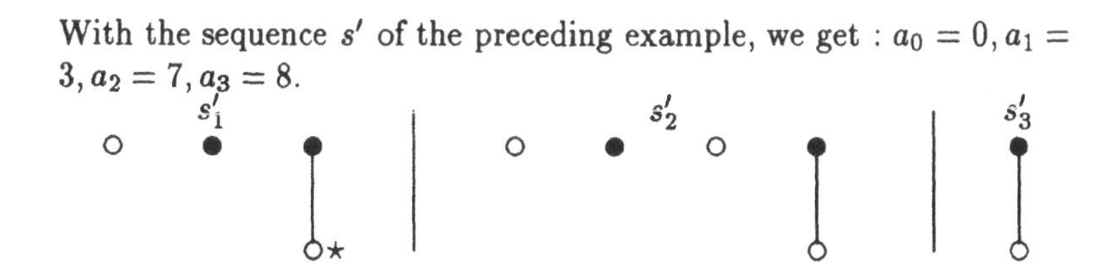

Now we define recursively a function f which maps a forest s'_i as follows :

read the sequence s'_i from left to right
if the tree read is ●
push ●
if the tree read T is ○ or
if the two last trees in the stack are a black and a white tree
pop these two trees and connect them to the white leaf of T
push the obtained tree
else
push T
the result is the forest that remains in the stack.

Example :

With the value of s'_2 of the preceding exemple, we get $f(s'_2)$:

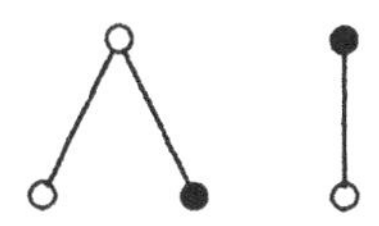

Definition 31 *Take $s \in W(p,q,k)$ and $s' = t_1(s)$. Then t_2 is the function which associates the list of forests $f(s'_1), \ldots, f(s'_{q-p})$ to s'.*

Example :

With the forest s' :

we get the list $s'' = t_2(s')$ defined by

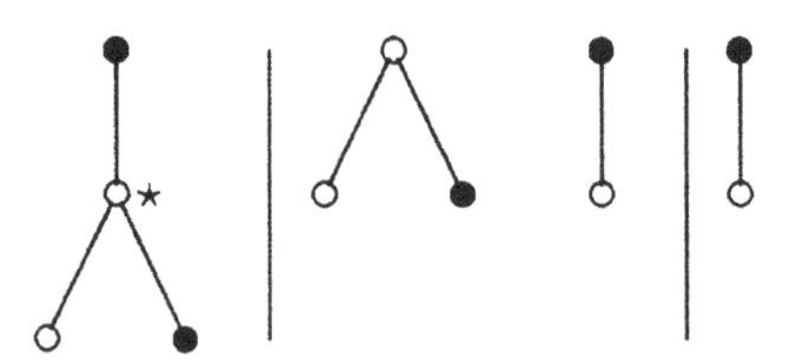

Now denote by s''_1 the forest $f(s'_1)$, ..., by s''_{q-p} the forest $f(s'_{q-p})$.

Theorem 13 *If $s \in W(p,q,k)$ and if $s'' = t_2(t_1(s))$ then :*

- *$\Phi(s'')$ is a list of $q-p$ forests $\Phi(s''_i)$, composed with $p+k$ black nodes and q white nodes.*
- *This list contains $q-p$ black trees that have at most two nodes and $q-p-k$ white trees.*
- *All forests $\Phi(s''_i)$ are formed with a black tree preceded by a certain number (eventually 0) of white trees; the root of this black tree is the only unary node of $\Phi(s''_i)$.*

Proof. The list $\Phi(s'')$ contains only back trees and white trees since the function f is able to build only such trees. In this list, like in s, we find $p+k$ black nodes and q white nodes.

Since each forest s'_i is composed with a single tree placed in last position, $f(s'_i)$ will have at least one black tree : the last tree of the forest. There will be only one because since each black leaf of $f(s'_i)$ is surrounded with two trees ○ or with the trees ○ and , all black leaves of s'_i will be popped by f and will be connected to the white leaf of the resulting tree. The black leaves of s'_i are leaves of $f(s'_i)$, therefore $s''_i = f(s'_i)$ contains only one unary node.

It remains to find the number of white trees belonging to $\Phi(s'')$. At the beginning we had p such trees, $2p - q + k$ of them have been poped with a black leaf and connected to a white node. The number being searched is therefore $p - 2p + q - k = q - p - k$.

□

We let a cyclic permutation act on the sequence s''. The construction of this permutation will be done in several steps :

- First, we map s'' to a word r formed by the letters x and y with the help of a function called g.
- Then we transform r into a 1-dominated sequence r' by replacing k of its letters y with x's and by applying a cyclic permutation.
- We add a label $+$ to the k trees of s'' which correspond through g to the k letters y which became letters x.
- Finally, we apply to this sequence the same cyclic permutation that was applied to r by g.

We get therefore a function t_3 which transforms a couple (s'', n) into a forest.

Let g be the function which replaces each tree T of the sequence s'' by :

- the letter y if T is the first tree in the sequence s''_i,
- or else by the letter x.

Example :

Returning to the sequence s'' obtained previously :

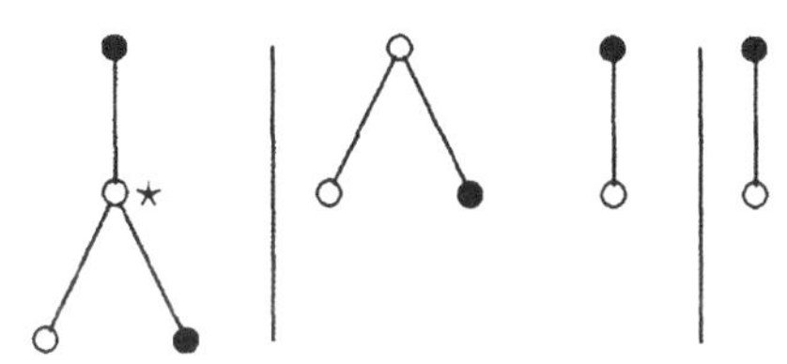

we get $g(s'') = yyxy$.

Theorem 14 *If $s \in W(p,q,k)$ and if $s'' = t_2(t_1(s))$ then $g(s'')$ is a sequence which contains $q-p$ letters y and $q-p-k$ letters x.*

Proof. According to Theorem 13, s'' is composed of $q-p$ sequences and contains $(q-p)+(q-p-k)$ trees. Since only $q-p$ trees are transformed into letters y, the $q-p-k$ remaining trees have to be transformed by g into letters x. □

Now we can define the height of a letter b in $g(s'')$ as the difference between the number of letters x and the number of letters y preceding b (b included). Let l be the minimal height of the letters y which we find in this sequence and let a_i be the first letter x of the sequence $g(s'')$ whose height is $l-i+k$. Since the height of the first symbol in the sequence is equal to ± 1 and that of the last symbol is equal to $-k$, the positions a_i are at least defined when $i \in [\![1, k]\!]$.

Definition 32 *Let $(s,n) \in W(p,q,k) \times [\![1,k]\!]$ and $s' = t_2(t_1(s))$, t_3 is then the function which associates to s' and to n the forest s'' which is formed :*

- *by adding the label $+$ to the trees of s' which appear in first position in a sequence s'_j and which are not in position a_i when i varies in $[\![1,k]\!]$,*
- *then by applying the cyclic permutation which transforms the a_n^{th} tree of s' into the first tree of the new forest.*

Example :

With the list of forests s' of the preceding section :

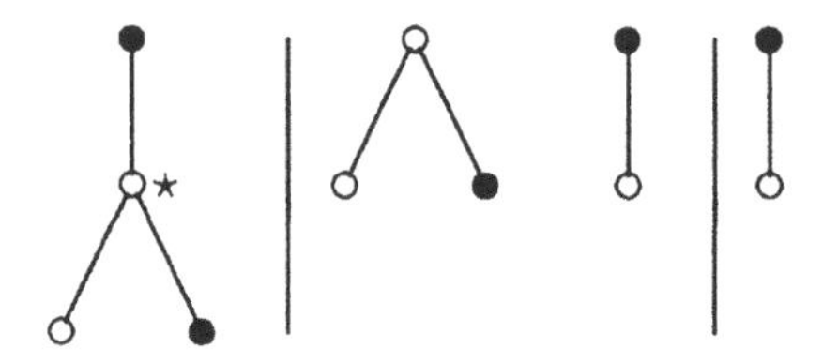

we have $g(s') = yyxy$, this gives $l = -2$ and $a_1 = 1$ and $a_2 = 2$. The forest $s'' = t_3(s', 2)$ is then :

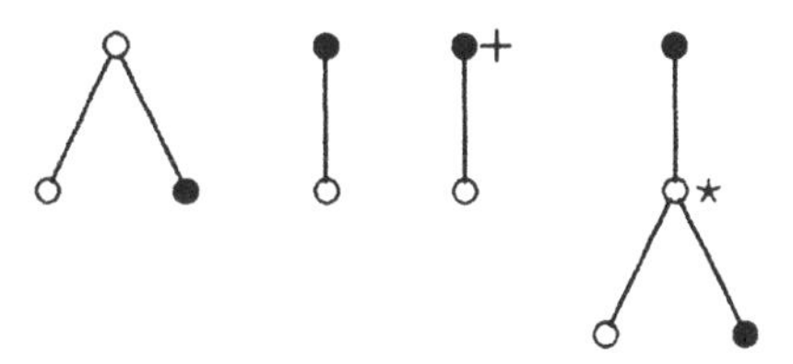

Let s'' be a forest of colored trees. We define now the function h that transforms the ith tree of s'' into the letter x if this tree has no label $+$ and into the letter y if it has label $+$. We get the following result :

Theorem 15 *If* $(s,n) \in W(p,q,k) \times [\![1,k]\!]$ *and if* $s'' = t_3(t_2(t_1(s)),n)$ *then* $h(s'')$ *is a 1-dominated sequence containing* $q-p$ *letters* x *and* $q-p-k$ *letters* y *(i.e. a letter* x *followed by a Dyck left factor).*

Proof. Let $s' = t_2(t_1(s))$. Of course, we have : $s'' = t_3(s',n)$. The trees of s'' with a label $+$ correspond to the beginning of a sequence s'_i of s' which was not in position $a_1, \ldots, a_k$; the set of these trees has cardinality $(q-p)-k$. Since these trees are transformed by h into the letter y, in $h(s'')$, we have $q-p-k$ letters y and $q-p$ letters x.

We need now to prove that $h(s'')$ is a 1-dominated sequence. So denote by r the word $g(s')$; we can decompose it as $r_1 l_1 r_2 l_2 \cdots r_k l_k r_{k+1}$ where $l_1, l_2, \ldots, l_k$ are the letters of r that appear in positions $a_1, a_2, \ldots, a_k$ as in the following :

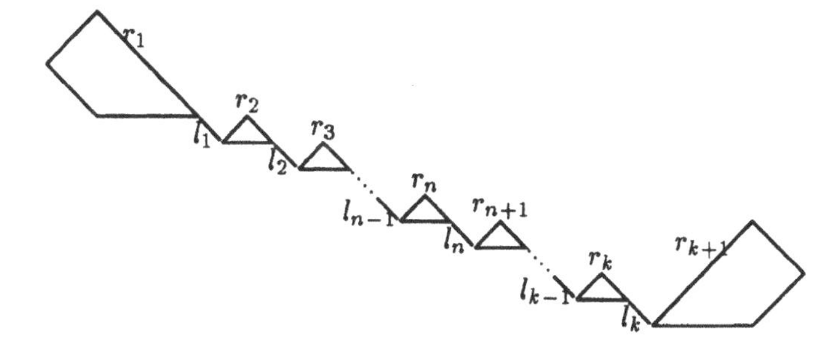

However we can transform this word r into $h(s'')$ by replacing the letters l_1, $\ldots$, l_k by a letter x and by applying the cyclic permutation wich transforms the a_n^{th} letter l_n into the first position. Therefore we obtain $h(s'')$ as

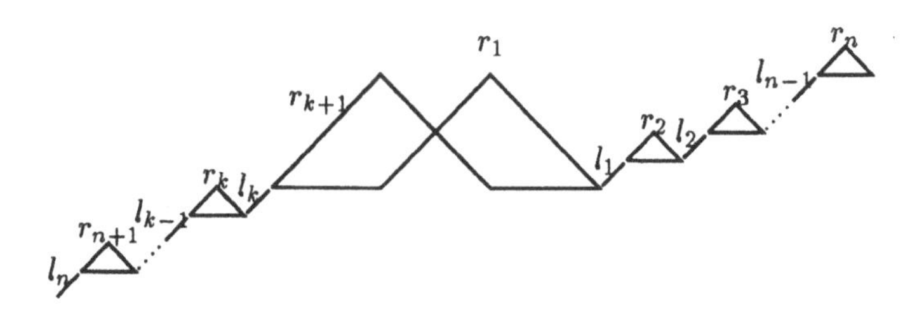

a word that is clearly a 1-dominated word. □

Now we can transform the obtained sequence into a forest of k trees with the help of the function t_4 defined recursively by the following procedure :

read the sequence s'' from left to right
if the tree T does not contain the label +
push T
if the tree T contains the label +
remove the label + from T
pop two trees
connect these trees to the leaf which is on the left edge of T
push the obtained tree
the resulting forest lies in the stack.

Example :

With the sequence s'' obtained previously :

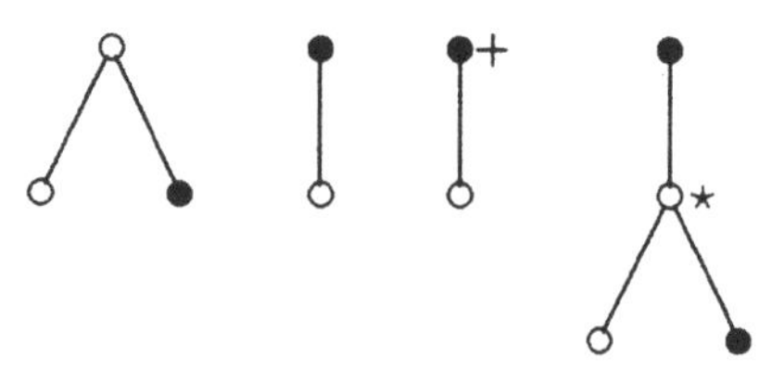

we get the forest :

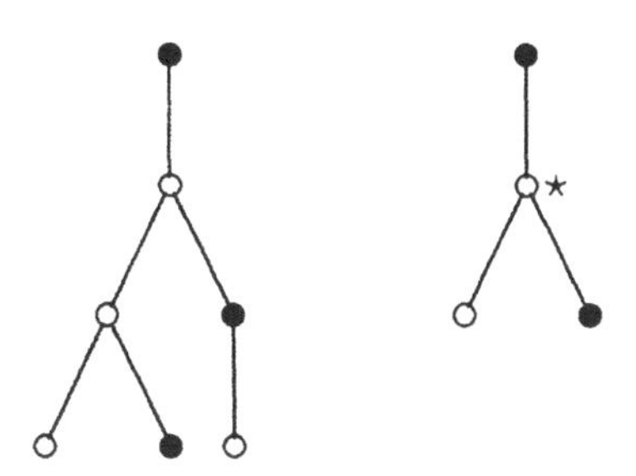

Proposition 12 *If* $(s,n) \in W(p,q,k) \times [\![1,k]\!]$ *and* $s'' = t_2(t_1(s))$, $s'' = t_3(s',n)$, *then the number of trees which are in the stack of* t_4 *after having read* j *trees is equal to the height of the* j^{th} *letter of the sequence* $h(s'')$.

Proof. This proof can be easily done by induction on j and it is left to the reader. □

This fact has two consequences. First, it validates the algorithm : since $h(s'')$ is a 1-dominated sequence, we are sure that we can find enough trees in the stack if we pop two trees and connect them to a black node. Secondly, this shows that the number of trees which will be in the stack is equal to k, the height of $h(s'')$'s last letter.

We can now prove the following theorem :

Theorem 16 *If* $(s,n) \in W(p,q,k) \times [\![1,k]\!]$ *then* $t_4(t_3(t_2(t_1(s)),n))$ *is a pointed forest with* k *black trees whose size is greater than or equal to* 2, *this forest contains* $p+k$ *black nodes and* q *white nodes.*

Proof. Let $s' = t_2(t_1(s))$, $s'' = t_3(s',n)$ and $s''' = t_4(s'')$. It remains to prove that the k trees of the obtained forest are black trees with at least two nodes.

We prove first that these trees are either black or white.

Lemma 3 *When we want to pop two trees to connect them to a leaf of a tree* T, *we find a black tree on the top of the stack.*

Proof. The tree T corresponds, in fact, to the first tree of a forest s'_i of $t_2(t_1((s,n)))$, the tree that is on the top of the stack corresponds therefore to the last tree of the preceding forest s'_{i-1} when $i > 1$ and s'_{q-p} when $i = 1$: this tree is always a black tree (cf. Theorem 13). □

Lemma 4 *When we want to pop two trees to connect them to a leaf of a tree T, the penultimate tree of the stack is a white tree.*

Proof. We must show that, if t_4 finds two trees T_1 and T_2 at the top of the stack and connects it to a white leaf, T_2 is a white tree. Let T be the smallest tree of s'' which will be transformed by t_4 into a subtree of T_1. This tree T has no label (+), else the tree of s'' which precedes T would be connected to a leaf of T and would belong to T_1, a contradiction to the choice of T. But T can't have a label + for two reasons :

- T was not the first tree of a forest s'_i of $t_2(t_1(s))$. In this case the left subtree of T in $t_2(t_1(s))$ was a white tree. T_2's root is the root of this subtree and is therefore a white node.
- this tree was the first tree of a forest s'_i of $t_2(t_1(s))$ but it was also in position a_i (for $i \in [\,1, k\,]$). Let F be the forest composed of the trees following T in s'' (T included). $h(F)$ is then a 1-dominated sequence. This means that T_2 will remain in the stack until the end of the procedure and will belong to the forest s''', a contradiction.

□

Theorem 13 shows that the sequence s'' contained at the beginning $2q - 2p - k$ trees, $q - p - k$ of them are black. At the end there are k trees, therefore $2q - 2p - 2k$ were used and connected to white nodes : $q - p - k$ trees are black and $q - p - k$ are white. The forest s''' has therefore $(q - p - k) - (q - p - k) = 0$ white trees and $(q - p) - (q - p - k) = k$ black trees.

In the same way, since we recover in s''' the black nodes and white nodes which we had in s, s''' contains $p + k$ black nodes and q white nodes. Theorem 13 proves that $t_2(t_1(s))$ does not contain a black tree with a single node. Therefore s'' and s''' can not have such node. The forest s''' has therefore k black trees whose size is greater or equal to two and contains $p + k$ black nodes and q white nodes. □

Now we define the function t_5 : t_5 transforms a forest of black trees whose size is greater than or equal to two, into a forest of white trees composed of the subtree children of the roots of the trees belonging to the initial set.

Example :

With the preceding forest s :

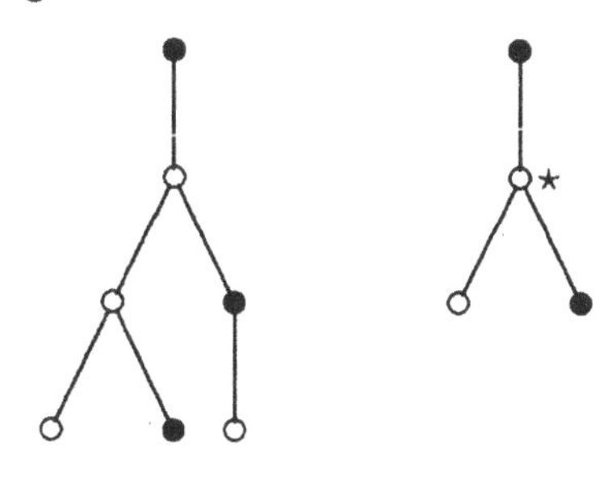

we get $t_5(s)$:

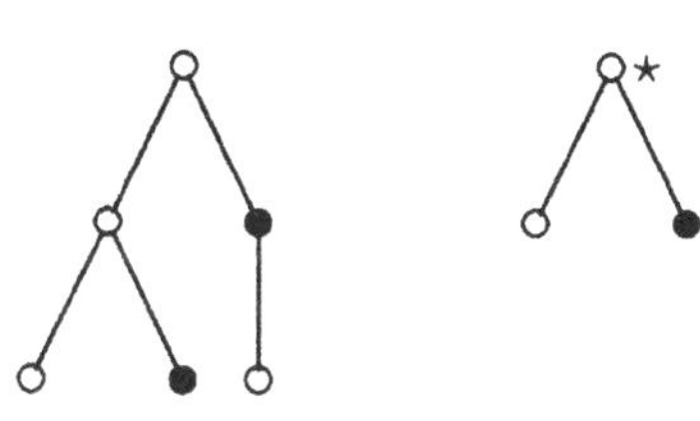

We define now t by $t((s,n)) = t_5(t_4(t_3(t_2(t_1)),n)))$, and the following theorem is a simple consequence of Theorem 16.

Theorem 17 *If* $(s,n) \in W(p,q,k) \times [\![1,k]\!]$ *then* $t((s,n)) \in N(p,q,k)$.

Proof. The function t_5 transforms a forest of k black trees into a forest of k colored trees by deleting k black nodes. Application of Theorem 16 ends the proof. □

Existence of t^{-1}

In the preceding section, we have defined a function t which maps an element $(s,n) \in W(p,q,k) \times [\![1,k]\!]$ to a pointed forest of colored trees of $N(p,q,k)$. We will prove below that this function is in fact a bijection. For this purpose, we will reverse the functions t_5, t_4, t_3, t_2, t_1 which define the function t.

Consider a pointed forest of k colored trees with p black nodes and q white nodes. It is clear that t_5^{-1} is the function which transforms a forest of k colored trees into a forest of k black trees formed by adding a black root to each of the k white trees.

Example :

If $t \in N(5, 10, 3)$ is the following forest :

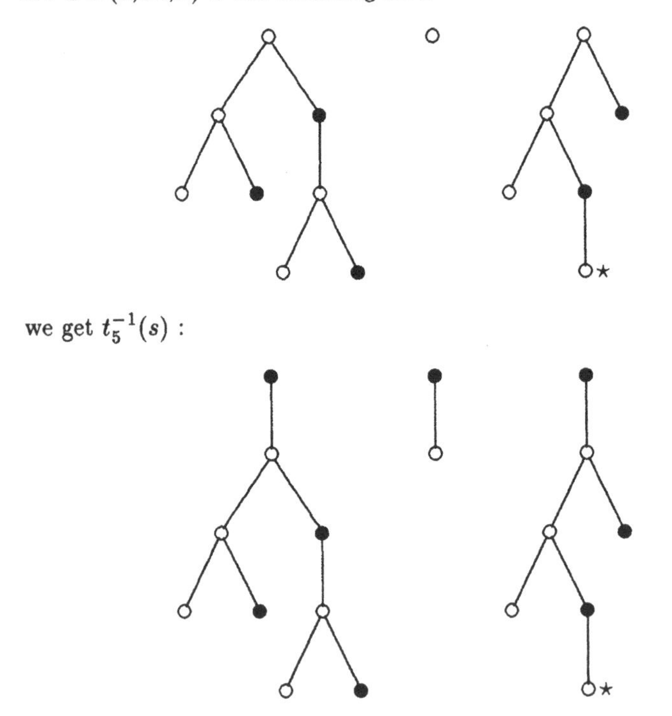

we get $t_5^{-1}(s)$:

We have the following theorem :

Theorem 18 *If $s \in N(p,q,k)$ then $t_5^{-1}(s)$ is the only forest s' of k black trees with at least two nodes such that $t_5(s') = s$.*

Proof. The only way to get a pointed forest F' of k black trees from a pointed forest F of k white trees, forest which gives F by removing the roots of the k black trees is by adding a black root to each of the k white trees of F. □

Let h' be the function which maps a tree T to a forest defined recursively as follows :

if T *is a leaf*
    $h'(T) = T$
if the root r *of* T *is a black unary node*
    let T_1 *be the subtree child of the root* r
    let $U_1, U_2, \ldots, U_u$ *be trees forming the forest* $h'(T_1)$
    $h'(T)$ *is the forest composed of the trees* $U_1, \ldots, U_{u-1}$
    and the tree with a black root whose child is U_u.
if the root of T *is a white binary node*
    let T_1 *(resp.* T_2*) be the left (resp. right) subtree of the root*
    if T_2 *is a leaf*
        let $U_1, U_2, \ldots, U_u$ *be the trees which compose the forest* $h'(T_1)$
        $h'(T)$ *is the forest whose trees are* $U_1, \ldots, U_{u-1}$ *and the tree with a white root whose left child is* U_u *and the right child a black leaf*
    else
        $h'(T)$ *is a forest composed of the trees of* $h'(T_1)$, *the trees of* $h'(T_2)$ *and the tree* $\circ +$ *(the tree* $\circ$ *with the label* $+$*).*

Let t_4^{-1} be the function which applies h' to all trees of the initial sequence and merges the obtained forests.

Example :

For example for the forest defined above we get

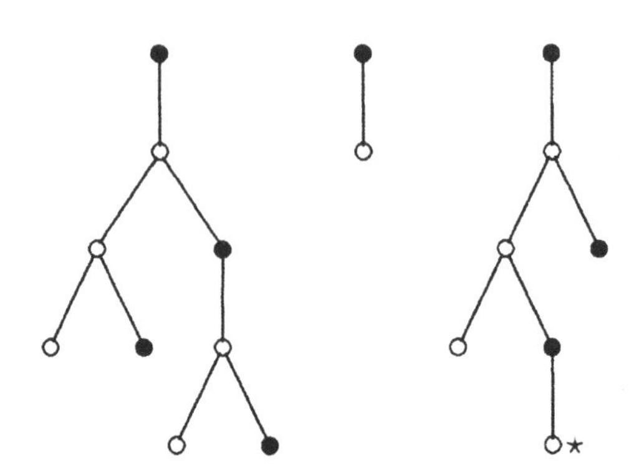

a forest $t_4^{-1}(s)$ which is equal to :

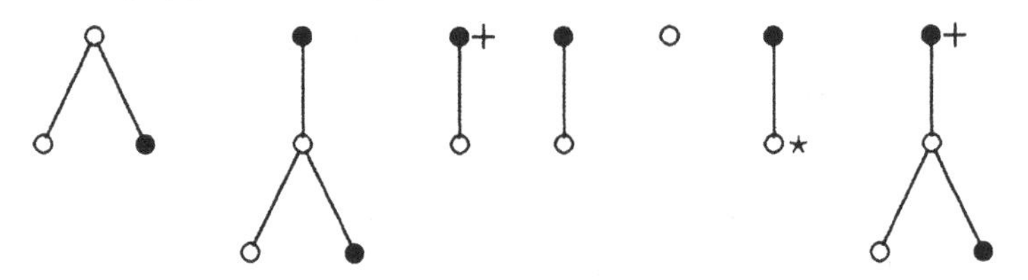

Theorem 19 *If $s \in N(p,q,k)$ and $s' = t_5^{-1}(s)$ then $s'' = t_4^{-1}(s')$ is the only sequence with the following properties :*

- *s'' does not contain a black tree reduced to a leaf,*
- *no node of s'''s trees is unary other than the root,*
- *s'' does not contain a pointed tree which is preceded by a white tree,*
- *$g(s'')$ is a 1-dominated sequence (g is the function defined page 93),*
- *$t_4(s'') = s'$.*

Proof. It is easy to see that if s' is composed of the trees $T_1, \ldots, T_k$ then s'' must be the forest obtained by concatenating forests $F_1', \ldots, F_k'$ where $t_4(F_i') = T_i$.

We proceed now by induction on the size of the tree T in order to prove that if T is a white tree or a black tree not reduced to a leaf, there exists only one forest F' which fulfills the properties $1-4$ of the theorem and $t_4(F') = T$.

In fact, if $T =\circ$, the only choice for F' is the forest composed of the tree $\circ$

Consider now a tree not reduced to a leaf.

- If T is a black tree whose child of the root is T_1, let $F'_1 = U_1, \cdots, U_u$ be the only forest which fulfills the assumptions of the theorem and $t_4(F'_1) = T_1$. Let U'_u be the black tree whose root has U_u as child. It is easy to see that $F' = U_1, \cdots, U_{u-1}, U'_u$ fulfills the properties $1-4$ of the theorem and $T = t_4(F')$. Assume now that there exists a second forest $F'' = V_1, V_2, \cdots, V_v$ with these properties and $T = t_4(F'')$. Since T is black, V_v must be black. This tree must contain at least two nodes since F'' does not contain a black tree reduced to a leaf. Let V'_v be the white subtree child of V_v's root. We have therefore a second forest $F''_1 = V_1, \ldots, V_{v-1}, V'_v$ which fulfills the properties $1-4$ of the theorem and $t_4(F''_1) = T_1$. This contradicts the induction hypothesis.

- If T is a white tree whose left child is a subtree T_1 and the right child a black leaf, the proof is the same as if T were a unary tree, just remark that the splitting $U_1, U_2, \ldots, U_u, \bullet$, $\circ$ $+$ is not possible since it would contain a black tree reduced to a leaf.

- If T is a white tree whose left child is a subtree of T_1 and the right child a subtree of T_2 not reduced to a leaf, we denote by F'_1 and F'_2 the only forests which verify the properties 1-4 of the theorem, $t_4(F'_1) = T_1$ and $t_4(F'_2) = T_2$. We can then prove that the sequence F' composed of F'_1's trees followed by F'_2's trees and $\circ$ $+$ fulfills our request.

  If now we have a second solution F'' formed with $U_1, \ldots, U_u$, the tree U_u must be a white tree.

  - If this tree has more than one node, its right child must be a black subtree V. This tree V must be $\bullet$, since U_u can only have one unary node at the root; $t_4(F'')$ will give a forest whose last tree has a white root. The right child of this root is a black leaf. This forest differs from T. We are lead to a contradiction.
  - If U_u is a white leaf, we find two forests F''_1 composed of the trees U_1, ..., U_l and F''_2 composed of the trees $U_{l+1}, \ldots, U_{u-1}$ which fulfill the properties 1-4 of the theorem to be proved, $t_4(U_1) = T_1$ and $t_4(U_2) = T_2$. The sequence F'' differs from the sequence of F'. This implies that F''_1 differs from F'_1 or that F''_2 differs from F'_2. We have therefore two forests which give T_1 or T_2. This contradicts the induction hypothesis.

□

Consider a forest $s \in N(p, q, k)$ and $s' = t_4^{-1}(t_5^{-1}(s))$.

The inverse function of t_3 removes first the labels $+$ in the forest s', then it splits this forest into a sequence of forests s'_i where s'_i is composed with the ith black tree of s' and the white trees which precede it. Finally, it applies the cyclic permutation which maps the sequence s'_i that has a white pointed leaf in the first position of a sequence s''.

We get therefore a sequence s''. Define the numbers a_i as on page 91 and n be the number such that the first tree of s' is in the a_n^{th} position of s''.

Example :

The preceding forest

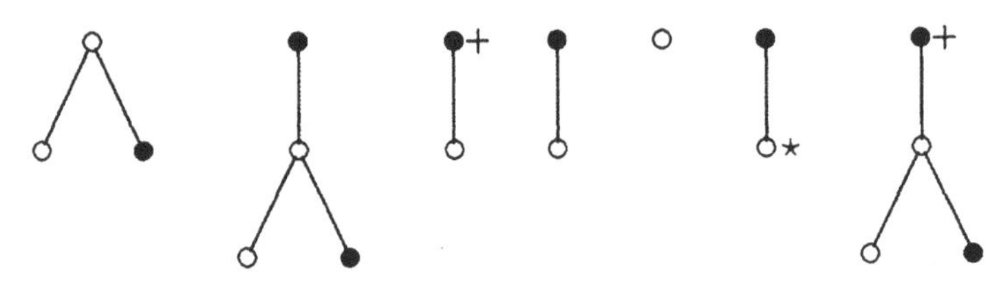

is transformed by t_3^{-1} into $(s', 2)$ with s' :

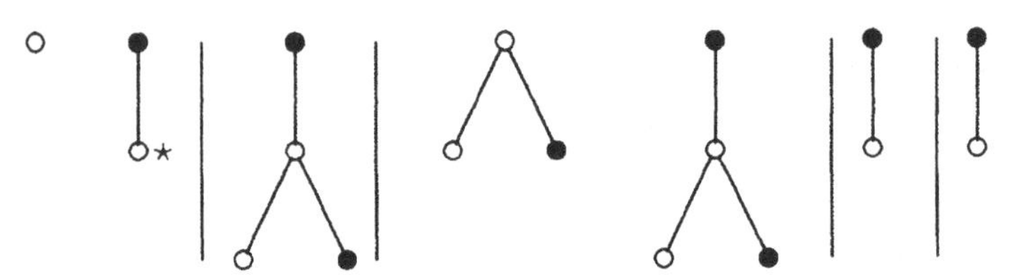

Theorem 20 *If $s \in N(p,q,k)$ and $s' = t_4^{-1}(t_5^{-1}(s))$ then $(s'', n) = t_3^{-1}((s', n))$ is the only pair such that*

- *s'' terminates with a black tree*
- *s'' has a starry node in the first sequence s''_1*
- *$t_3((s'', n)) = s'$.*

Proof. It is easy to see that the sequence s'' can be obtained in only one way: there is a unique cyclic permutation which transforms the sequence s' into a sequence which ends with a black tree and which has a starry tree which is not preceded by any white trees.

Now we have to determine the value of the integer n. For doing so, it is enough to use the position of the first tree of s' in s''. □

For the inverse function of t_2, apply the transformation f' defined below to each tree T of the initial sequence :

if T a leaf
  we return T
if T is a unary node
  if the child of T is a leaf, we return
  if the child of T has two children T_1 and T_2,
    we return the forest formed with the trees of $f'(T_1)$ followed by those of $f'(T_2)$ and by .
if T is a binary node with children T_1 and T_2
  we return the forest formed with the trees of $f'(T_1)$ followed by these of $f'(T_2)$ and by ○ .

Example :

With the list s' introduced previously :

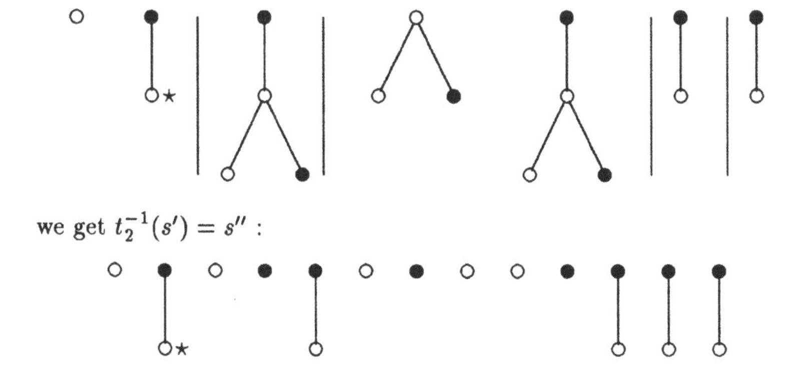

we get $t_2^{-1}(s') = s''$:

Theorem 21 *If $s \in N(p,q,k)$, $(s',n) = t_3^{-1}(t_4^{-1}(t_5^{-1}(s)))$ then $s'' = t_2^{-1}(s')$ is the only couple such that*

- $t_2((s'', n)) = (s', n)$,
- *s'' is only composed with trees* ● , ○ , $\overset{\bullet}{\circ}$,
- *each black leaf of s'' is preceded by a white tree.*

Proof. We use the same technique as for the proof of Theorem 19. We prove first that the problem of defining the value of $t_2^{-1}(F)$ for a forest F reduces to the problem of defining $t_2^{-1}(T_i)$ for each tree T_i of the forest F, then we prove by induction that there exists only one forest F' of trees ○ and $\overset{\bullet}{\circ}$ such that $t_2(F') = F$.

Consider a tree T:

- if T is equal to ○ or $\overset{\bullet}{\circ}$, the forest $F' = (T)$ is the only forest which satisfies $(T, n) = t_2((F', n))$,
- if T is the tree with a white root whose left subtree is T_1 and whose right subtree is ● , we can use for F' the forest composed of the trees of $t_2^{-1}(T_1)$ followed by ● and ○ . The existence of another solution would imply the existence of a second forest F_1'' such that $t_2(F_1'') = T_1$, and this would contradict the induction hypothesis.
- if T is a black tree, T has at least two nodes. Let T_1 be the child of T's root and $F_1' = U_1, \ldots, U_{v-1}$, ○ the forest such that $t_2(F_1') = T_1$. The forest $F' = U_1, \ldots, U_{v-1}$, $\overset{\bullet}{\circ}$ is then the only forest such that $t_2(F') = T$, the existence of another solution would, as previously, imply the existence of a second forest F_1'' such that $t_2(F_1'') = T_1$.

The splitting is therefore unique. In addition, since the black trees reduced to a leaf appear by splitting a white tree which had a black leaf as right child and a white tree as left child, all black leaves of s'' will be preceded by a white node.

□

It remains to show that t_1^{-1} is the cyclic permutation which places in first position the tree with a white pointed leaf and then removes the label $\star$ from this leaf.

Example :

The forest s

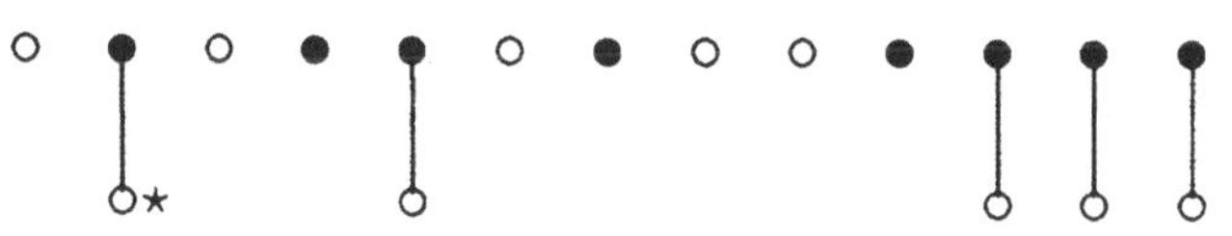

is transformed by t_1^{-1} into s' :

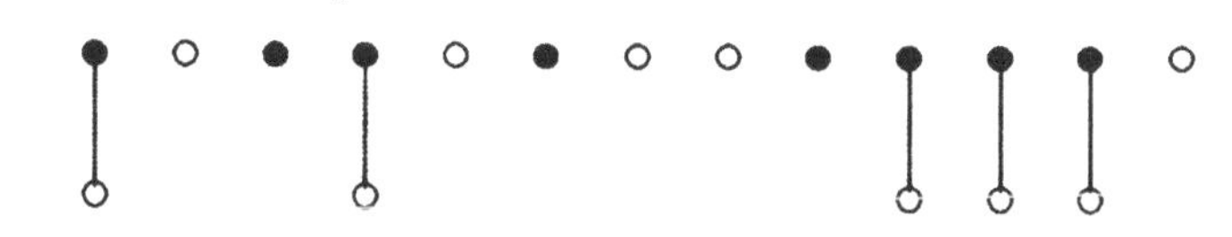

Theorem 22 *If $s \in N(p,q,k)$, $(s',n) = t_3^{-1}(t_4^{-1}(t_5^{-1}(s)))$ and $s'' = t_2^{-1}(s')$ then $s''' = t_1^{-1}(s'')$ is the only element of $W(p,q,k)$ such that $t_1(s''') = s''$.*

Proof. The function t_1 adds the label $\star$ to the white leaf which is in the first tree of the forest which is transformed by it. Then t_1 realizes the cyclic permutation which takes the last tree $\overset{\bullet}{\underset{\circ}{|}}$ and puts it in the last position. Therefore, we use the permutation which brings the pointed tree of s'' in first position, then we can remove the label $\star$ from this tree. Since the cyclic permutation is defined uniquely by this condition, we get a good definition for t_1^{-1}.

Now we must prove that the sequence s''' belongs to $W(p,q,k)$.

Theorem 21 states that s'' is only composed of trees $\circ$, $\bullet$, $\overset{\bullet}{\underset{\circ}{|}}$, the black trees are always preceded by a white tree. The forest s''' is obtained through a cyclic permutation of s'', therefore s''' is also a forest composed with trees $\circ$, $\bullet$, $\overset{\bullet}{\underset{\circ}{|}}$ in which all black trees $\bullet$ are preceded by a white tree. The sequence s'' contains as many trees $\bullet$, $\overset{\bullet}{\underset{\circ}{|}}$ as $t_5^{-1}(s'')$ contains black leaves and unary nodes. It contains therefore $2p - q + k$ trees $\bullet$ and $q - p$ trees $\overset{\bullet}{\underset{\circ}{|}}$. s'' contains as many white nodes as $t_5^{-1}(s)$, this gives a number of trees $\circ$ equal to $q - (q - p) = p$ (indeed this is equal to the number of white nodes which do not belong to a tree $\overset{\bullet}{\underset{\circ}{|}}$). □

6.4 GENERATION OF COLORED TREES

The generation of a forest of k colored trees is now easily done in two steps: the random selection of an element s of $W(p,q,k)$ with uniform probability and of a random integer n in $[\![1, k]\!]$. Then we compute the value of $t((s,n))$ in linear time with the help of the functions t_1, t_2, t_3, t_4 and t_5 and finally we remove the star $\star$ from the pointed node in the constructed forest. In doing so, we get each forest of k colored trees with p black nodes and q white nodes with probability:

$$\frac{q}{k\ card(W(,q,k))}.$$

It remains only to see how to generate randomly an element of $W(p,q,k)$. This is done by distributing $p+k$ black nodes in the q places which follow the q white nodes. We get $\binom{q}{p+k}$ different sequences with uniform probability.

Then we choose $q-p$ black nodes between the $p+k$ which we just placed, and we replace these $q-p$ black nodes and the $q-p$ leaves which precede them by $q-p$ trees ⚲ .

Theorem 23 *We obtain all elements of $W(p,q,k)$ and each of them is generated with probability:*

$$\frac{1}{\binom{q}{p+k}\binom{p+k}{q-p}}.$$

Proof. By construction, all obtained forests are composed with $q-(q-p)$ ○ , $q-p$ ⚲ , $p+k-(q-p)$ ● such that each black leaf ● is preceded by a white tree ○ . We obtain therefore only elements of $W(p,q,k)$. It remains now to prove that we get all elements of $W(p,q,k)$ uniquely.

Take $s \in W(p,q,k)$ and replace the trees ⚲ of s by two trees: the tree ○ followed by ● . Then we get a forest with $p+k$ black leaves and q white leaves which can be uniquely obtained by placing the $p+k$ black trees in the q positions which follow a white leaf. With this forest, we obtain s by choosing the $q-p$ black nodes which have appeared in the decomposition of the trees ⚲ and by replacing it and the white leaf which precedes it by a tree ⚲ . This is the only way to get the forest s. □

Corollary 6 *The number of forests of k colored trees with p black nodes and q white nodes is equal to*

$$\frac{k}{q}\binom{q}{p+k}\binom{p+k}{q-p}.$$

Proof. Theorem 23 indicates that $card([\![1,k]\!] \times W(p,q,k))$ is equal to $A = k\binom{q}{p+k}\binom{p+k}{q-p}$. There exist therefore A pointed trees in $N(p,q,k)$. This corresponds to $\frac{A}{q}$ forests of k colored trees which are composed of p black nodes and q white nodes. □

6.5 CONCLUSION

We have presented a linear time algorithm for the generation of forests with k colored trees, p black nodes and q white nodes. This family of trees can not be generated directly with the techniques developed for binary trees, trees split in patterns, etc... We have seen in the last chapters how to generate some trees in linear time by using 1-1 correspondences. We will see in the next chapter a different approach that gives algorithms whose average complexity is linear.

7

TWO METHODS OF GENERATION BY REJECTION

Abstract

We present two very simple and efficient algorithms which built randomly and uniformly a unary-binary tree and a skew tree of size n. The average complexity of these algorithms is $O(n)$. These algorithms differ greatly from the algorithms which we have seen in the previous chapters. In fact, they are based on the idea that we can choose an enlarged probability space where uniformity can be achieved and then filter out the desired trees without destroying uniformity.

7.1 INTRODUCTION

We propose in this chapter two methods of generation by rejection to generate a unary-binary tree and a skew tree. These methods are very simple and efficient; indeed, they have an average complexity in $O(n)$. However, the complexity in the worst case is infinity. This fact is intrinsicaly tied to the methods of generation by rejection. These methods are based on the following principle : Begin the generation of one structure and if in the procedure we generate a worst structure, we stop and we go back to the beginning.

But even if this problem exists, this method allows us to find very efficient algorithms for the generation of unary-binary trees and skew trees when only slower deterministic algorithms are known.

There are some 1-1 correspondences between the unary-binary trees and the Motzkin words. Skew trees are in 1-1 correspondence with another kind of words

: the Motzkin left factors. We present algorithms to builds a Motzkin word and a Motzkin left factor and we explain how the 1-1 correspondences work.

Such algorithms have important applications in percolation theory. Left Motzkin factors are in 1-1 correspondence with combinatorial objects called directed animals [41] that have been widely used for modelling the physical phenomenon of percolation [29].

7.2 FIRST METHOD

This section is based on an algorithm of Barcucci et al [13] that build a Motzkin left factor of size n. We explain how this algorithm works, then we show how this algorithm can be used to generate a unary-binary tree or a skew tree.

7.2.1 Generation of a Motzkin left factor

This very clever algorithm was presented by E. Barcucci, R. Pisani, R. Sprugnoli [13] in the beginning of the nineties. It has a loop that draws k consecutive letters x, y or a with probabilities $\frac{1}{3}$. This loop stops when $k = n$ or when the word that was generated has more letters y than letters x.

In the first case when $k = n$, we have uniformly obtained a Motzkin left factor. In the second case, the generation has failed and we begin a new drawing from the beginning.

It is easy to see that this algorithm draws each Motzkin left factor with a uniform probability. Indeed, each time the loop ends, the probability that a *special* Motzkin left factor m is drawn is exactly $\frac{1}{3^n}$ (the probability to draw one by one the n letters of the Motzkin left factor m). When we use many loops, the probabilities of drawing different Motzkin left factors remain equal, thus the algorithm works as claimed.

7.2.2 Complexity

We show now that the average complexity of this algorithm is in $O(n)$. For this purpose, let note $k(n)$ the time that we expect to pass in a loop when we begin to draw a Motzkin left factor of size n, and let P_n be the number of Motzkin left factors of size n.

Theorem 24 *We have*

$$k(n) \sim 2\frac{\sqrt{3}}{\sqrt{\pi}}\sqrt{n}.$$

Proof. Indeed, the probability of drawing a ith letter is equal to the ratio between the number of Motzkin left factor of size $i-1$ and 3^{i-1} (the number of different ways to choose $i-1$ letters x, a, y). Thus the average complexity of one drawing is

$$P_0 + \frac{P_1}{3} + \cdots + \frac{P_{n-1}}{3^{n-1}}$$

which is asymptotically equivalent to

$$2\frac{\sqrt{3}}{\sqrt{\pi}}\sqrt{n}$$

using $P_n \sim \frac{\sqrt{3}}{\sqrt{\pi n}}3^n$ (see [51]) when n tends to infinity. □

Now we just need to discover that the probability that one drawing succeeds is

$$R = \frac{P_n}{3^n} \sim \frac{\sqrt{3}}{\sqrt{\pi n}},$$

and that the probability to do the ith drawing is $(1-R)^{i-1}$.

The average complexity of this algorithm is :

$$k(n)\left(1 + (1-R) + (1-R)^2 + \cdots + (1-R)^i + \cdots\right) = \frac{k(n)}{R} \sim 2n.$$

So this algorithm is linear.

This algorithm can now be used to generate two kinds of trees, the unary-binary trees and the skew trees.

7.2.3 Generation of a unary-binary tree : the Marty Algorithm

We show now how to use the generation of a Motzkin left factor in order to obtain an algorithm that generates uniformely a unary-binary tree, this algorithm is inspired by many works of X.G. Viennot and is dedicated to Marion Marty.

First, we define a new of words, Marty words, by

Definition 33 *The Marty words are the words defined by a grammar M_a such that*

$$M_a = \epsilon + xMyM_a$$

where M is the grammar of the Motzkin words

$$M = \epsilon + aM + xMyM$$

(see Definition 5 page 12).

In fact, the Marty words are the Motzkin words that do not have a letter a with height 0. For instance $xyxaxaaxyyy$ is a Marty word.

The generation of a Motzkin word is done in four steps :

- first, we generate a Motzkin left factor p of size $n+1$ that is not a Marty word,
- we map this word p to a Motzkin left factor p' with an odd final height,
- then, we map p' to a word m with $n+1$ letters x, y, a such that $|m|_x = |m|_y + 1$,
- finally, we use the cycle lemma to transform this word into a Motzkin word.

The first step is really simple; indeed we need only to generate a Motzkin left factor with size $n+1$ until we find a Motzkin left factor that is not a Marty word. The average complexity of this step is $O(n)$ because the probability to draw a Marty word is only in $O(\frac{1}{n})$. (It is less than the probability to draw a Motzkin word that is $\frac{M_{n+1}}{P_{n+1}} = O(\frac{1}{n})$).

The last step is a linear step that we have already seen (see the Lemma 2 page 76). So we need only to see how to do the two intermediary steps in linear time. This approach is in the same spirit as that of X.G.Viennot [51].

Transformation of a Motzkin left factor that is not a Marty word into a Motzkin left factor with an odd final height

Let take a Motzkin left factor that is not a Marty word. Two situations can arise

- either p is already a Motzkin left factor with an odd final height; of course we take p' equal to p
- or the height of p is even. In this case, we define a mapping f that transforms p into a word p' by :

Definition 34 *The function f maps a word p to a word p'*

- *In replacing the first letter a of p with height 0 by a letter x, if such a letter exists.*
- *In replacing the last letter x of p with height 1 by a letter a, if the previous transformation is impossible.*

Example :

If we use the mapping f on a word p

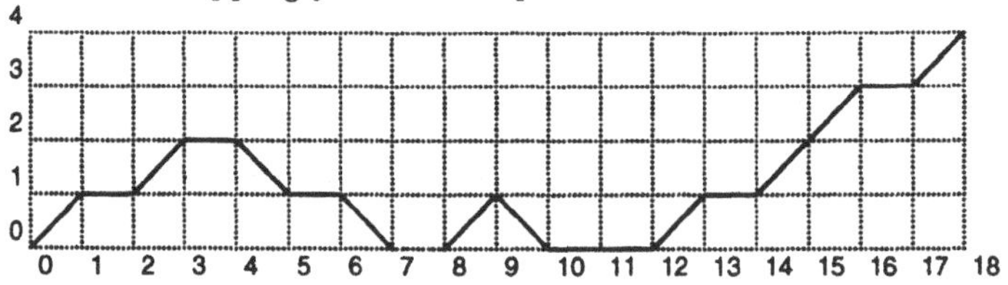

we find a word p'

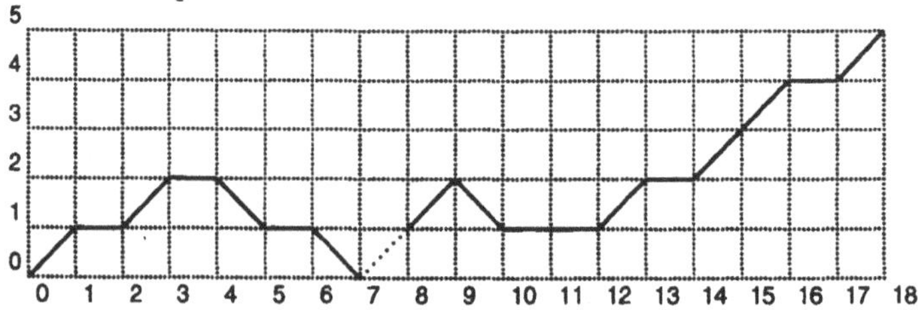

This function is a 1-1 correspondence that can be computed in time $O(n)$. Indeed, we have $f(f(w)) = w$ for each Motzkin left factor that is not a Marty word.

Transformation of a Motzkin left factor with an odd final height into a word m such that $|m|_x = |m|_y + 1$.

We can now define another 1-1 correspondence g that maps a Motzkin left factor p' with an odd final height into a word m such that $|m|_x + |m|_y + |m|_a = n+1$ and $|m|_x = |m|_y + 1$.

Definition 35 *Let g be the mapping that maps a Motzkin left factor p' with size $2h+1$ by translating the last letters x of p' with height h' into a letter y for each h' in $[\,1, h\,]$.*

Example :

The word p'

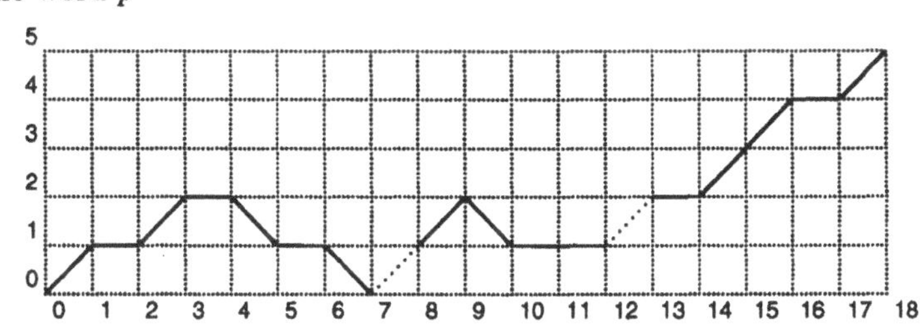

is mapped by g to m

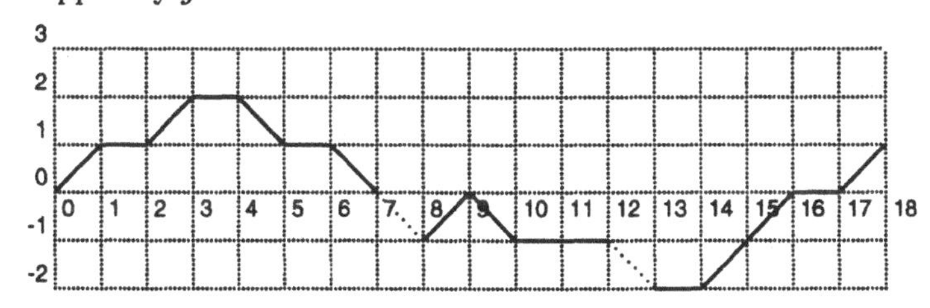

It is easy to see that g maps a Motzkin left factor to a word m with $n+1$ letters x, y, a that has one more letters x than letters y.

This mapping is a 1-1 correspondence. Indeed, if we take a word m with $n+1$ letters x, y, a that has one more letters x than letters y and if we replace in this word the first letters y that have height less than 0 by a letter x we obtain again the word p'.

The complexity

The four steps have an average complexity in $O(n)$, so does this algorithm. We need now to show only that it generates all Motzkin words with uniform probability.

Theorem 25 *Each Motzkin word of size n is generated with a probability* $\frac{2(n+1)}{P_{n+1}-M_{a\,n+1}} = \frac{1}{M_n}$ *where M_{a_n} is the number of Marty words with size $n+1$.*

démonstration :

Indeed, let take a Motzkin word m' with size n. There are $n+1$ words m with $n+1$ letters x, a, y that map to $x.m'$ using a cyclic permutation.

Each of these $n+1$ words has unique antecedent p' by g . Each of these $n+1$ antecedents has an unique antecedent by f (remember that f and g are two 1-1 correspondences) and another antecedent by the identity mapping. These $2(n+1)$ words are Motzkin left factors that are not Marty words.

So, the probability to generate the word m' is equal to the sum of the probabilities to generate the $2(n+1)$ Motzkin left factors with which they are in 1-1 correspondence. Thus it is equal to : $2(n+1)\frac{1}{P_{n+1}-M_{a\,n+1}}$.

The drawing is uniform.

□

We can now transform the obtained Motzkin word m into an unary-binary tree. Indeed, if we replace each letter x, y and a by the letters x, xyy and xy we obtain a new word m'. Then the word xm' which is a 1-dominated word can be transformed into a unary-binary tree using the 1-1 correspondence g defined on page 39.

Example :

The word $xxaxxaxayayyxyaaya$ is mapped to the word m' :

$$xxxyxxxyxxyxyyxyxyyxyyxxyyxyxyxyyxy.$$

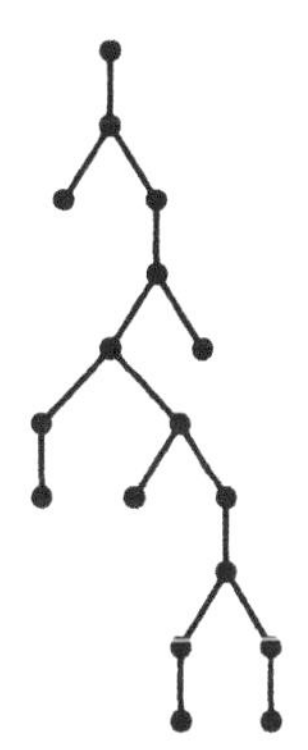

7.2.4 Generation of a skew tree: Penaud's Algorithm

We present here a new class of trees : the skew tree. These trees have been defined by J.G. Penaud and are very interesting because they are in 1-1 correspondence with the Motzkin words and some directed animals (see [41]).

Definition 36 *Let us take a tree T whose nodes can have a left child or/and a right child. If v is a node of T, we denote by displacement(v) the difference between the number of right and left edges in the path from the root of T to v. Then we say that*

- *A skew tree is a tree T such that for every binary node x in T there is no node z in the right subtree of the tree defined by x such that the displacement of z in the subtree defined by x is less or equal to zero.*
- *A square tree is a skew tree T such that for each node v in T, the displacement of v is greater or equal to zero.*
- *A sharp tree is a square tree T such that a leaf v of T exists such that displacement(v) is equal to zero.*

Example :

For instance, if we take these three trees,

the first is a skew tree but is not a square tree, the second is a square tree but is not a sharp tree and the last one is a sharp tree.

Now we have the following theorem that was first proved by J.G. Penaud.

Theorem 26 *There exist* 1-1 *correspondences between Marty words, Motzkin words, Motzkin left factors of size* $n-1$ *and respectively sharp trees, square trees and skew trees with* n *nodes.*

Proof. Here we show that the definitions of these two differents structures are in correspondence.

In the following, we denote

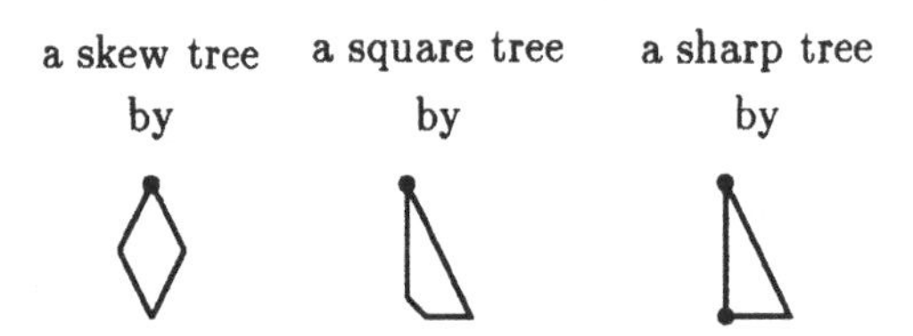

Now, we see that the grammatical definition of Motzkin left factor and a skew tree are in correspondence :

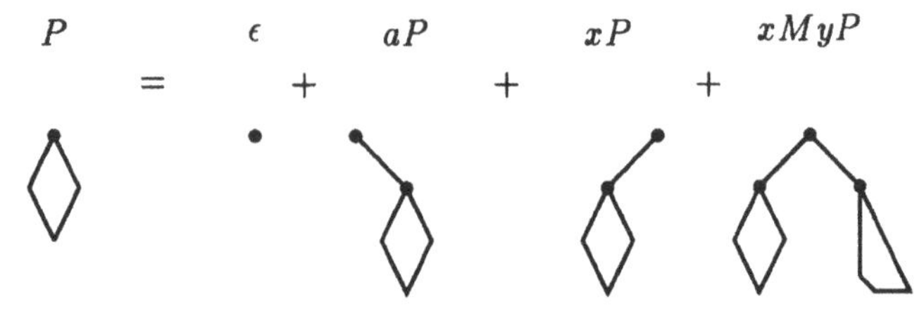

(i.e. a Motzkin left factor is the empty word, a letter a or x followed by a Motzkin left factor or a letter x followed by a Motzkin word then by a letter

y and finally by a Motzkin left factor. Similarly a skew tree is a leaf, or a root that has a skew tree as left or right child, or a root that has a skew tree as left child and a square tree as right child).

The same is true for the Motzkin word and the square tree :

$$M = M_a + M_a a M$$

(i.e. a Motzkin word is a Marty word or a Marty word followed by a letter a and by a Motzkin word. Similarly a square tree is a sharp tree or a sharp tree whose leaf v such that $displacement(v) = 0$ has a square tree as right child),

and also for the Marty word and the sharp tree :

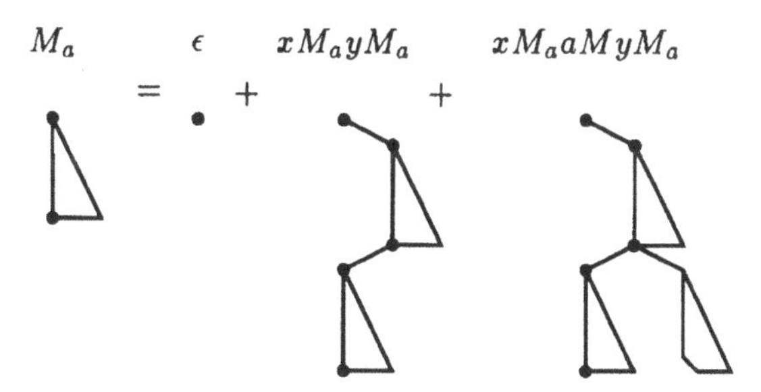

□

Therefore, to generate a skew tree of size n, we begin to generate a Motzkin left factor w of size $n-1$. Then we replace in w the last letters x that has height h' by a letter b for each h' in $[\![1, height(w)]\!]$. Finally, we use the following algorithm that is based on the preceding proof; it is written in a pseudo-C language.

```
Tree *new_node ()

        Tree *node = malloc (sizeof (Tree))
        node.left = node.right = node.thread = nil
        return node


Tree *read_P ()
        Tree *node = new_node ()
        if peek_next_letter () != '' then // this is not the last letter
          switch read_next_letter ()
                  case 'a' :
                        node.right = read_P ()
                        break
                  case 'b' :
                        node.left = read_P ()
                        break
                  case 'x' :
                        node.right = read_M ()
                        new_letter () // the letter y
                        node.left = read_P ()
        return node


Tree *read_M ()
        Tree *node
        node = read_Ma ()
        if peek_next_letter () == 'a' then
          node.thread.right = read_M ()
        return node


Tree *read_Ma ()
        Tree *node
        node = new_node ()
        if peek_next_letter () == 'x' then
          read_next_letter ()
          node.right = read_Ma ()
          if read_next_letter () == 'a' then
            node.thread.right = read_M ()
            read_next_letter () // the letter y
          node.thread = (node.thread.left = read_Ma ()).thread
        else
          node.thread = node
        return node
```

To read the letters of w from the left to the right, we have used two functions; the first *peek_next_letter* returns the letter of w that we are watching, the second *read_next_letter* returns the same value, and constrains us to watch the next letter.

And we have represented a node v by a structure of three elements : the two first ones called *left* and *right* give the left and the right of that node; the last one called *thread* gives the leaf w of a node v such that $displacement(w) = 0$ when the subtree defined by v is a sharp tree.

7.3 ANOTHER METHOD BY REJECTION

We present here in detail another algorithm to generate some Motzkin words in average time $O(n)$; this algorithm is dedicated to Samaj Lareida. We see later that the same method can be also used to generate Motzkin word. These methods can be used to generate efficiently words on a parallel computer. This is not true for the two algorithms given in the beginning of the chapter.

First we present the algorithm of Samaj Lareida to generate a Motzkin word and thus a unary-binary tree. In the first part, this algorithm chooses the number k of letters x. In the second part, it builds a random Motzkin word that has k letters x out of n letters : this construction can be done in linear time using the tools which we have described in the subsection 5.5.5 (page 83). Therefore, we focus only on the first part. It is straightforward to write an algorithm to generate the number of letters x in a random Motzkin word of length n using $O(n)$ arithmetic operations; unfortunately these algorithms need large preprocessing time to compute some coefficients and they manipulate integers of great size so their real complexity in bit operations is large. This algorithm is simple, it has an average complexity in $O(n)$ that remains in $O(n\, log(n))$ when we count bit operations.

First, we define the numbers $M_{n,l}$ and $N_{n,l}$. The numbers $M_{n,l}$ enumerate the Motzkin words having $l-1$ letters x out of n letters, and the $N_{n,l}$ are certain numbers satisfying :

$$\forall n, \forall l, M_{n,l} \leq N_{n,l}$$

We end that subsection making some remarks about the numbers $M_{n,l}$ and $N_{n,l}$.

Then we present our algorithm. The principles of this algorithm are very simple: The basic idea is to choose a slightly enlarged probability space where uniformity can be achieved and then to filter out the desired words without destroying uniformity. More precisely, in order to generate a Motzkin word of size n, we take $\lceil \frac{2n+2}{3} \rceil + 1$ ballot-boxes, and for $l \in [0, \lceil \frac{2n+2}{3} \rceil]$ we put $N_{n,l}$ words in the lth ballot-box : the $M_{n,l}$ Motzkin words that have $l-1$ letters x out of n letters and $(N_{n,l} - M_{n,l})$ other words. Then we choose a random ballot-box k with probability :

$$\frac{N_{n,k}}{\sum_{l=0}^{\lceil \frac{2n+2}{3} \rceil} N_{n,l}},$$

we choose a random word in the kth ballot-box and see if this word is a Motzkin word or not. If it *is*, the first part of the algorithm finishes, we can draw a Motzkin word that has $k-1$ letters x out of n letters. Otherwise, we apply again the first part of the algorithm.

We will see that this algorithm can be easily implemented with the definition of $N_{n,l}$ given below and that the average complexity of the algorithm is $O(n)$.

7.3.1 Definitions and basic facts

Let $M_{n,l+1}$ denote the number of Motzkin words with l letters x out of n letters.

We recall that

Proposition 13 ▪ *For all* l, $1 \leq l \leq \lfloor \frac{n}{2} \rfloor + 1$, *we have* $M_{n,l} = \frac{1}{n+1} \binom{n+1}{l, l-1, n-2l+2}$,

- *for all* l, $l \leq 0$ *or* $l > \lfloor \frac{n}{2} \rfloor + 1$, *we have* $M_{n,l} = 0$,
- *and as* n *goes to infinity*

$$\sum_{l=1}^{\lfloor \frac{n}{2} \rfloor + 1} M_{n,l} \sim \frac{\sqrt{3}\, 3^{n+1}}{2\sqrt{\pi} n^{\frac{3}{2}}}$$

Proof. The properties 1-2 of this proposition follow directly from the Formula 5.1 (page 83). The proof of the asymptotic characterization of $\sum_l M_{n,l}$ can be found in [51]. □

We define now numbers $N_{n,l}$ by

$$N_{n,l} = \frac{1}{n+1} \binom{n+1}{\lfloor \frac{n+1}{3} \rfloor} \binom{\lceil \frac{2n+2}{3} \rceil}{l}.$$

Remark :

In the above and further formulas, we use the fact that $n - \lfloor \frac{n}{3} \rfloor = \lceil \frac{2n}{3} \rceil$.

Then we have the following proposition that allows us to compare the values of $M_{n,l}$ and $N_{n,l}$ and motivates the choice of the numbers $N_{n,l}$:

Proposition 14 *For all l, $1 \leq l \leq \lfloor \frac{n}{2} \rfloor + 1$, we have :*

$$\frac{M_{n,l}}{N_{n,l}} = \frac{\binom{a_n}{c_n}}{\binom{b_n}{c_n}}$$

with

- *if* $l \leq \lfloor \frac{n+1}{3} \rfloor$, $a_n = \lfloor \frac{n+1}{3} \rfloor$, $b_n = n - 2l + 2$, $c_n = \lfloor \frac{n+1}{3} \rfloor - l + 1$,
- *if* $l > \lfloor \frac{n+1}{3} \rfloor$, $a_n = \lceil \frac{2n+2}{3} \rceil - l$, $b_n = l - 1$, $c_n = l - \lfloor \frac{n+1}{3} \rfloor - 1$.

Proof. Using the values of $M_{n,l}$ and $N_{n,l}$, we find :

$$\frac{M_{n,l}}{N_{n,l}} = \frac{n!}{l!(l-1)!(n-2l+2)!} \frac{\lfloor \frac{n+1}{3} \rfloor! l! (\lceil \frac{2n+2}{3} \rceil - l)!}{n!} = \frac{\lfloor \frac{n+1}{3} \rfloor!}{(l-1)!} \frac{(\lceil \frac{2n+2}{3} \rceil - l)!}{(n-2l+2)!}.$$

This gives, when $l \leq \lfloor \frac{n+1}{3} \rfloor$,

$$\frac{M_{n,l}}{N_{n,l}} = \frac{\binom{\lfloor \frac{n+1}{3} \rfloor}{\lfloor \frac{n+1}{3} \rfloor - l + 1}}{\binom{n - 2l + 2}{\lfloor \frac{n+1}{3} \rfloor - l + 1}}.$$

When $l > \lfloor \frac{n+1}{3} \rfloor$, the proof is similar. □

This proposition has an interesting corollary.

Corollary 7

$$\textit{For all } l,\ 1 \leq l \leq \left\lfloor \frac{n}{2} \right\rfloor + 1, \textit{ we have } M_{n,l} \leq N_{n,l}.$$

Proof. We need only to prove that when

- $l \leq \lfloor \frac{n+1}{3} \rfloor$ then $\lfloor \frac{n+1}{3} \rfloor \leq n + 2 - 2l$,
- $l = \lfloor \frac{n+1}{3} \rfloor + 1$ then $M_{n,l} = N_{n,l}$
- $l > \lfloor \frac{n+1}{3} \rfloor + 1$ then $\lceil \frac{2n+2}{3} \rceil - l \leq l - 1$

which is straightforward. □

Finally, we have a proposition that compares the sum of the $M_{n,l}$'s with the sum of the $N_{n,l}$'s :

Proposition 15

$$\frac{\sum_{l=0}^{\lceil \frac{2n+2}{3} \rceil} M_{n,l}}{\sum_{l=0}^{\lceil \frac{2n+2}{3} \rceil} N_{n,l}} \sim \frac{1}{\sqrt{3}}.$$

Proof. Using Proposition 13, we have :

$$\frac{\sum_{l=0}^{\lceil \frac{2n+2}{3} \rceil} M_{n,l}}{\sum_{l=0}^{\lceil \frac{2n+2}{3} \rceil} N_{n,l}} \sim \frac{\sqrt{3}\, 3^{n+1}}{2\sqrt{\pi} n^{\frac{3}{2}}} \frac{n}{2^{\lceil \frac{2n+2}{3} \rceil} \binom{n+1}{\lfloor \frac{n+1}{3} \rfloor}}.$$

The result follows from Stirling's formula $n! \sim \sqrt{2\pi n} \left(\frac{n}{e}\right)^n$ ([30] page 467). □

7.3.2 Principles of this algorithm

We review in this subsection the principles of the algorithm. First, we take $\lceil \frac{2n+2}{3} \rceil + 1$ ballot-boxes labelled 0, 1, ..., $\lceil \frac{2n+2}{3} \rceil$. Then we put in each ballot-box l $N_{n,l}$ words : the $M_{n,l}$ Motzkin words that have $l - 1$ letters x out of n letters and $(N_{n,l} - M_{n,l})$ other words (recall that $M_{n,l} \leq N_{n,l}$).

Then we choose randomly a ballot-box k with probability

$$\frac{N_{n,k}}{\sum_{l=0}^{\lceil \frac{2n+2}{3} \rceil} N_{n,l}}.$$

Finally, we accept or reject this choice k, depending on a random outcome. We accept this choice with probability

$$\frac{M_{n,k}}{N_{n,k}}.$$

If this test fails, we go back to the beginning. But if this test succeeds, we draw a random Motzkin word with $k-1$ letters x out of n letters, this can be done easily in time $O(n)$ using Subsection 5.5.5 (page 83).

We have the following result that proves the uniformity of the algorithm.

Theorem 27 *This algorithm builds each Motzkin word of size n with probability*

$$\frac{1}{\sum_{l=1}^{\lfloor \frac{n}{2} \rfloor + 1} M_{n,l}}.$$

Proof. First, suppose that if the test of k's correctness fails, the algorithm stops. Then this new algorithm will return no result with probability

$$\alpha_n = 1 - \frac{\sum_{k=0}^{\lceil \frac{2n+2}{3} \rceil} M_{n,k}}{\sum_{l=0}^{\lceil \frac{2n+2}{3} \rceil} N_{n,l}}$$

and will build a Motzkin word m with probability

$$\frac{N_{n,k}}{\sum_{l=0}^{\lceil \frac{2n+2}{3} \rceil} N_{n,l}} \frac{M_{n,k}}{N_{n,k}} \frac{1}{M_{n,k}} = \frac{1}{\sum_{l=0}^{\lceil \frac{2n+2}{3} \rceil} N_{n,l}}$$

where $k-1$ is the number of letters x of m. (Indeed, we have chosen the ballot-box k with probability $N_{n,k}/\sum_{l=0}^{\lceil \frac{2n+2}{3} \rceil} N_{n,l}$, a Motzkin word in this ballot-box with probability $M_{n,k}/N_{n,k}$, and the Motzkin word m in the set of all Motzkin words with n letters and $k-1$ letters x with probability $1/M_{n,k}$).

So, when the algorithm succeeds, it builds a Motzkin word of size n with uniform probability.

Now, if we study the real algorithm that goes back to the beginning when the test of k's correctness fails, the probability that we choose a Motzkin word m of size n, becomes

$$\begin{aligned}\frac{1}{\sum_{l=0}^{\lceil\frac{2n+2}{3}\rceil} N_{n,l}}(1+\alpha_n+\alpha_n^2+\ldots) &= \frac{1}{\sum_{l=0}^{\lceil\frac{2n+2}{3}\rceil} N_{n,l}}\frac{1}{1-\alpha_n} = \frac{1}{\sum_{l=0}^{\lceil\frac{2n+2}{3}\rceil} M_{n,l}} \\ &= \frac{1}{\sum_{l=1}^{\lfloor\frac{n}{2}\rfloor+1} M_{n,l}}.\end{aligned}$$

□

7.3.3 The Algorithm of Samaj Lareida

Using the values of $N_{n,l}$ of the first section, the implementation becomes straightforward.

In order to generate a random Motzkin word of size n :

- we need first to choose a ballot-box k with probability :

$$\frac{N_{n,k}}{\sum_{l=0}^{\lceil\frac{2n+2}{3}\rceil} N_{n,l}} = \frac{\binom{\lceil\frac{2n+2}{3}\rceil}{k}}{2^{\lceil\frac{2n+2}{3}\rceil}}.$$

  This can be done by generating $\lceil\frac{2n+2}{3}\rceil$ random bits (0 or 1) and by calling k the number of 1 bits that have been generated.

- Then we need to choose a word in the kth ballot-box and test if this word is or is not a Motzkin word. Using Proposition 14 and Corollary 7, the probability that the chosen word is a Motzkin word, is equal to

$$\frac{M_{n,k}}{N_{n,k}} = \frac{\binom{a}{c}}{\binom{b}{c}}$$

  where a, b, c are some integers that satisfy $b \geq a \geq c$. This probability is equal to the probability that a placement of c objects in b positions corresponds to a placement of c objects in the a first positions.

So, we compute first the values of a, b and c using Proposition 14, then we draw a random placement of c objects in b positions and we check for the c objects in the first a positions. If the answer is positive, we generate a Motzkin word that has $k-1$ letters x out of n letters. If the answer is negative, we go back to the beginning.

This gives us the following algorithm written in a "pseudo-C" language :

do
    // choice of the kth *ballot-box*
    generate $\lceil \frac{2n+2}{3} \rceil$ *bits and put in* k *the number of* 1 *bits*
    // validation of this choice
    if $k \leq \lfloor \frac{n+1}{3} \rfloor$
        $a = \lfloor \frac{n+1}{3} \rfloor, \ b = n - 2k + 2, \ c = \lfloor \frac{n+1}{3} \rfloor - k + 1$
    else
        $a = \lceil \frac{2n+2}{3} \rceil - k, \ b = k - 1, \ c = k - \lfloor \frac{n+1}{3} \rfloor - 1$
    flag $=$ *true*
    if $k = 0$ *or* $k > \lfloor \frac{n}{2} \rfloor + 1$
        flag $=$ *false*
    else
        while $c > 0$
            $N = 1 + random(b)$ *// a random number in* $[\![1, b]\!]$
            if $N > a$
                flag $=$ *false*
                break
            else
                $c = c - 1$
                $a = a - 1$
                $b = b - 1$
while flag $=$ *false*
generate a random Motzkin word with $k - 1$ *letters* x *out of* n *letters.*

7.3.4 The complexity

The aim of the section is to prove the following theorem :

Theorem 28 *The average complexity of the algorithm is in* $O(n)$.

Proof. First, we need to remark that the worst case complexity C_n of the choice of k and the validation of this choice is in $O(n)$.

Then let α_n be the probability that the choice of k is not valid when we try to generate a Motzkin work of size n. We get :

$$\alpha_n = 1 - \frac{\sum_{k=0}^{\lceil \frac{2n+2}{3} \rceil} M_{n,k}}{\sum_{l=0}^{\lceil \frac{2n+2}{3} \rceil} N_{n,l}}.$$

Therefore, using Proposition 15, we have as n goes to infinity

$$\alpha_n \sim 1 - \frac{1}{\sqrt{3}}.$$

The proof ends by noting that the average complexity of the algorithm is

$$O((C_n + \alpha_n C_n + \ldots + \alpha_n^l C_n + \ldots) + n) = O\left(\frac{n}{1-\alpha_n} + n\right) = O(n)$$

To obtain the first expression above, note that we must first choose an initial value for k and test if this value is correct, which takes time in $O(C_n)$; the choice of k and its validation are done an $(l+1)$st time with probability α_n^l, $l = 1, 2, \cdots$. Finally, we always build a Motzkin word that has $k-1$ letters x out of n letters, and the complexity of this step is $O(n)$. □

7.3.5 Another application to generate skew trees: Denise's Algorithm

This algorithm was first presented by A. Denise in [18]. We proceed in the same way as for the generation of a Motzkin word. First we choose k the number of letters x and y of our Motzkin left factor words. Then we generate a Motzkin left factor with k letters a.

We use the following theorem which presents a 1-1 correspondence "à la Viennot" between the words with $\lceil \frac{k}{2} \rceil$ letters x, $\lfloor \frac{k}{2} \rfloor$ letters y and $n-k$ letters a and the Motzkin left factor with k letters x and y and $n-k$ letters a.

Theorem 29 *There is a 1-1 correspondence between the words that have* $\lceil \frac{k}{2} \rceil$ *letters* x, $\lfloor \frac{k}{2} \rfloor$ *letters* y *and* $n-k$ *letters* a *and the Motzkin left factor with* k *letters* x *and* y *and* $n-k$ *letters* a.

Proof. Let us take a word v that has $\lceil \frac{k}{2} \rceil$ letters x, $\lfloor \frac{k}{2} \rfloor$ letters y and $n-k$ letters a.

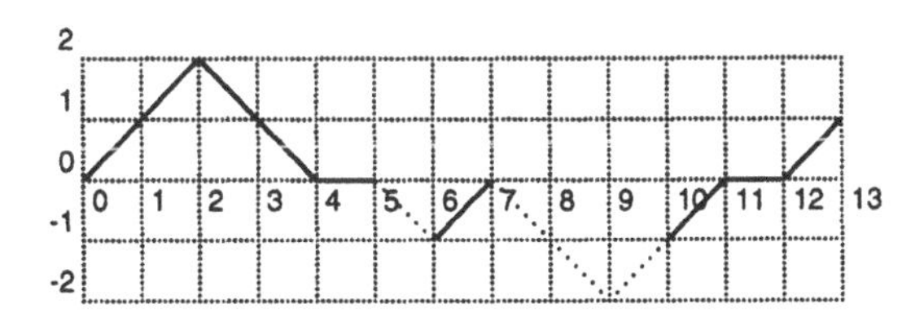

Then we replace in v the letters x (resp. y) that have height $j < 0$ by a letter y (resp. x).

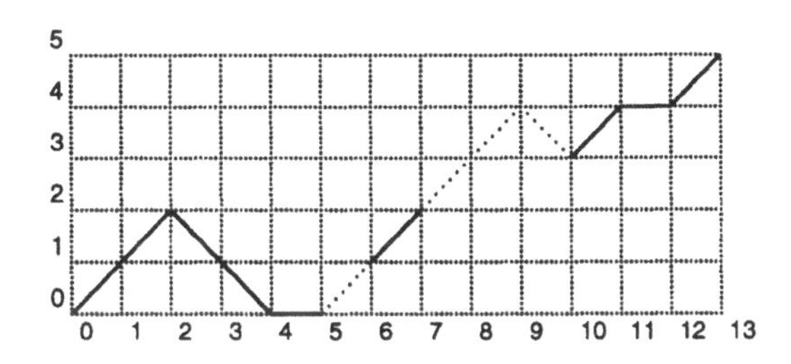

We have obtained a Motzkin left factor w that has $n-k$ letters a and k letters x and y. This transformation is a 1-1 correspondence. Indeed let us denote by h the height of w and by w_0, w_1, ..., w_h the $h+1$ Motzkin words such that $w = w_0 l_1 w_1 \cdots l_h w_h$ where l_1, ..., l_h are some letters x as follows

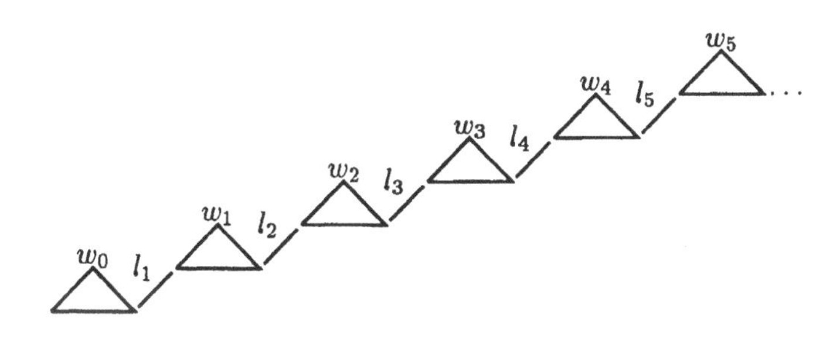

then if we replace respectively the letters x and y in l_1w_1, l_3w_3, ..., $l_{2\lfloor\frac{h-1}{2}\rfloor-1}w_{2\lfloor\frac{h-1}{2}\rfloor-1}$ by a letter y and x, we find again the word v as follows

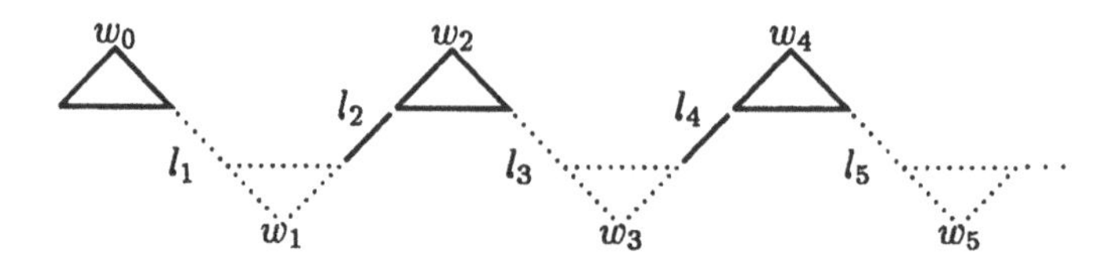

□

Corollary 8 *The number of Motzkin left words that have k letters x and y and $n-k$ letters a is*

$$M_{n,k} = \binom{n}{\lfloor\frac{k}{2}\rfloor, \lceil\frac{k}{2}\rceil, n-k}.$$

Proof. This is the number of ways to mix $\lfloor\frac{k}{2}\rfloor$ letters x, $\lceil\frac{k}{2}\rceil$ letters y and $n-k$ letters a. □

Using this 1-1 correspondence, it is easy to build a Motzkin left word that has k letters x and y and $n-k$ letters a. However, before we can begin this generation we must first choose the number k, we proceed as for the generation of a Motzkin word.

We begin by defining a number $N_{n,k}$ by

Definition 37 *For each k' in $[\,0, \lceil\frac{2n}{3}\rceil\,]$, we define*

$$N_{n,2k'-1} = N_{n,2k'} = \binom{n}{\lfloor\frac{n}{3}\rfloor, k', \lceil\frac{2n}{3}\rceil - k'}$$

We have now the following proposition :

Proposition 16 *For all k such that $k > n$ or $k < 0$, we have $\frac{M_{n,k}}{N_{n,k}} = 0$.*

For all k such that $0 \leq k \leq n$, denote $k' = \lfloor \frac{n}{2} \rfloor$ and $\epsilon = \{\frac{n}{2}\}$, we have

$$\frac{M_{n,k}}{N_{n,k}} = \frac{\binom{a}{c}}{\binom{b}{c}} \leq 1$$

with

- *If $0 \leq k' \leq \lfloor \frac{n}{3} \rfloor$ then $a = \lfloor \frac{n}{3} \rfloor$, $b = n - \epsilon - 2k'$ and $c = \lfloor \frac{n}{3} \rfloor - k'$.*
- *If $\lfloor \frac{n}{3} \rfloor < k' \leq \lfloor \frac{n-\epsilon}{2} \rfloor$ then $a = \lceil \frac{2n}{3} \rceil - k' - \epsilon$, $b = k'$ and $c = k' - \lfloor \frac{n}{3} \rfloor$.*

Furthermore, we have

$$\frac{\sum_{k=0}^{n} M_{n,k}}{\sum_{k=-1}^{2\lceil \frac{2n}{3} \rceil} N_{n,k}} \sim \frac{1}{\sqrt{3}}$$

Proof. The proof of this proposition is similar to the proof of Propositions 14,15 and of Corollary 7. □

Therefore we obtain the following linear algorithm to build a skew tree :

First, we find the number of letters a in our Motzkin left factor.

- We generate $\lceil \frac{2n}{3} \rceil$ random bits (0 or 1) and call k' the number of bits 1. Then we generate another random bit ϵ.
- We accept or reject the choices k' and ϵ, depending on a random outcome as we have done for a Motzkin word.
  - If $k' > \lfloor \frac{n-\epsilon}{2} \rfloor$, we reject these choices.
  - If $0 \leq k' \leq \lfloor \frac{n}{3} \rfloor$, we accept these choices with probability $\binom{a}{c} / \binom{b}{c}$ where $a = \lfloor \frac{n}{3} \rfloor$, $b = n - \epsilon - 2k'$, $c = \lfloor \frac{n}{3} \rfloor - k'$,
  - If $\lfloor \frac{n-\epsilon}{2} \rfloor \geq k' > \lfloor \frac{n}{3} \rfloor$, we accept the choices of k' and ϵ with probability $\binom{a}{c} / \binom{b}{c}$ where $a = \lceil \frac{2n}{3} \rceil - k' - \epsilon$, $b = k'$, $c = k' - \lfloor \frac{n}{3} \rfloor$.

  If the choice of the pair (k', ϵ) is rejected, we go back to the beginning.

Then, we generate a Motzkin left factor with $k = 2k' + \epsilon$ letters x and y. We proceed as follows. First we generate a permutation of n elements. Then we replace the values of this permutation that are in $[\![1, \lceil \frac{k}{2} \rceil]\!]$ by x, those that are in $[\![\lceil \frac{k}{2} \rceil + 1, k]\!]$ by y and the remaining values by a. Finally, we compute the height of each letter and transform a letter $u_i = x$ (resp. $u_i = y$) into y (resp. x) if and only if u_i has a negative height.

Finally, we use the 1-1 correspondence between skew trees and Motzkin left factors to obtain a skew tree with uniform probability.

7.4 CONCLUSION

We have presented many algorithms that generate uniformly a unary-binary tree or a skew tree of size n. The average complexity of these algorithms is in $O(n)$, and their maximum complexity is infinite.

This last fact is a disadvantage of the methods by rejection which can be easily outcome. Indeed, if we have an algorithm which generates our structure which has a worst case complexity equal to $C(n)$, we use first a method by rejection and if after the $C(n)$ first steps, we have not drawn any tree, we stop with the method of generation by rejection the next time an attempt of generation fails and we use the deterministic algorithm.

In our cases, if we have $n = o(C(n))$, we get a uniform algorithm whose average complexity remains in $O(n)$ and the worst case complexity in $C(n)$.

8

ARBORESCENCES

Abstract

We show in this chapter how to generate recursively some arborescences. These methods explain, for example, how we can use some combinatorial results to obtain a generation algorithm.

8.1 INTRODUCTION

The random generation of arborescences does not fit into the scope of the techniques developed in the previous chapters. For the first time, we do not know any useful 1-1 correspondence between these trees and some words that can be easily generated. Therefore, we use another approach. We will try to generate recursively one of these trees. We will obtain an algorithm which was presented for the first time in [38] and which was called by its authors RANRUT (an abbreviation of RANdom Rooted Uniform Tree).

This approach is very interesting because it can be used to generate many classes of trees [27] but the corresponding algorithms have a complexity that is often larger than when there exists a simple 1-1 correspondence.

The organization is as follows : Section 2 defines arborescences and gives some useful results while Section 3 presents the RANRUT algorithm. The average case analysis comprises the main part of Section 4. We examine first the complexity of the preprocessing step, then we consider two situations :

- case 1 : all arithmetic operations have a complexity in $O(1)$.

- case 2 : the real complexity of these operations is accounted for.

The study of the real complexity of these operations is important because this algorithm manipulates some very large numbers.

In Section 5, we give another example of a recursive algorithm which builds a random binary arborescence with n nodes and Section 6 summarizes the results and concludes the chapter.

8.2 ARBORESCENCES, DEFINITIONS AND PROPERTIES

First, we recall the definition of an arborescence.

Definition 38 *An arborescence on the finite set $S \neq \emptyset$ is a pair $(r, \{T_1, \ldots, T_k\})$ (with $k \geq 0$) such that*

- *r is an element of S,*
- *T_1, ..., T_k are arborescences defined on the sets S_1, ..., S_k,*
- *the set $\{r\}$, S_1, ..., S_k form a partition of S.*

We call node the elements of S, root of T the node r and we say that k edges exist that connect the node r to the root of the arborescences T_1, ..., T_k.

Furthermore, we consider that the two arborescences T and T' are the same if there exists an isomorphism from S to S' which transforms T into T'.

To draw an arborescence T in this chapter, we will order the edges child of each node of T. The chosen ordering is of course arbitrary and there exist many pictures which represent the same arborescence.

Example :

Two pictures of the same arborescence:

Some pictures of the four arborescences that have four nodes :

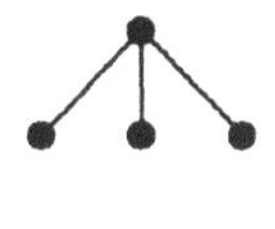   

Denote by t_n the number of arborescences with n nodes and by f the generating function defined by :

$$f(z) = \sum_{i=1}^{\infty} t_i z^i.$$

Property 3 *We have :*

1. $t_1 = 1,\ t_2 = 1,\ t_3 = 2,\ t_4 = 4, \ldots,$
2. $t_n = \frac{1}{n-1} \sum_{d=1}^{n-1} \sum_{j=1}^{\lfloor \frac{n-1}{d} \rfloor} d\, t_{n-jd} t_d,$
3. $f(z) = z\ exp\left(f(z) + \frac{1}{2} f(z^2) + \frac{1}{3} f(z^3) + \cdots + \frac{1}{n} f(z^n) + \cdots\right),$
4. $t_n \sim c\frac{\rho^n}{n^{\frac{3}{2}}}$ *with* $c = 0.4399\ldots$ *et* $\rho = 2.9559\ldots$.

démonstration :

The first property is obvious. The third one derives from classical results about generating functions that we will skip in this chapter (see [16] for more details). The fourth one is actually the most difficult, it can be proven by studying the singularities of f in the complex plane. The paper of Meir and Moon [37] gives the details of this proof.

Here, we will only prove the second property since the algorithm RAN-RUT is in fact based on it.

First, suppose that we have chosen a coherent representation of all the arborescences, ie. a representation that is such that : "if we take a tree T, denote by r its root and by $T_1, \ldots, T_k$ the children subtrees of r then the representations of the arborescences $T_1, \ldots, T_k$ appear in the representation of T".

Exemple :

The representation of these three arborescences satisfies this property :

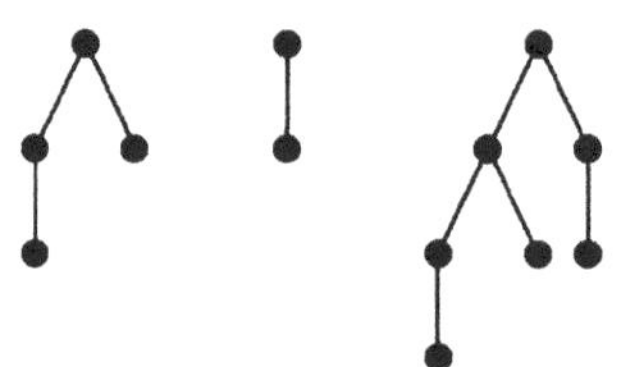

This one is not a coherent representation :

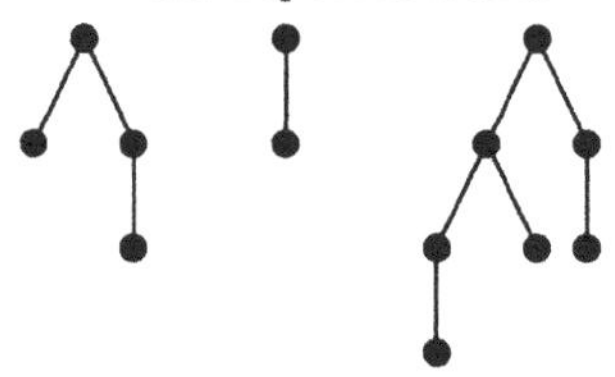

Of course, a coherent representation of all the arborescences can be built recursively and there exist many such representations.

But let us go back to our problem. We want to show

$$(n-1)t_n = \sum_{d=1}^{n-1} \sum_{j=1}^{\lfloor \frac{n-1}{d} \rfloor} d\, t_{n-jd} t_d.$$

The first term of this equality $(n-1)t_n$ is in fact the number of arborescences T with n nodes in which a node different from the root of T has been chosen.

Take an arborescence T with n nodes and denote by r the root of T and v a node of T that is different from r. We can now use our coherent representation to draw the tree T and denote by $T_1, \ldots, T_k$ the arborescences, children of r that we find in this picture from the left to the right.

Now, denote by

- T_l rooted child of r that contains the node v,
- d the number of nodes of T_l,

- j the number of arborescences T_m with $m \in [\, l, k]$ such that $T_m = T_l$.

When we disconnect these j subtrees T_l from the root of T, we obtain a new arborescence T' with $n - jd$ nodes and j different occurrences of T_l, one of these containing the node v.

Example :

An arborescence representation of T where the node v is enclosed in a circle

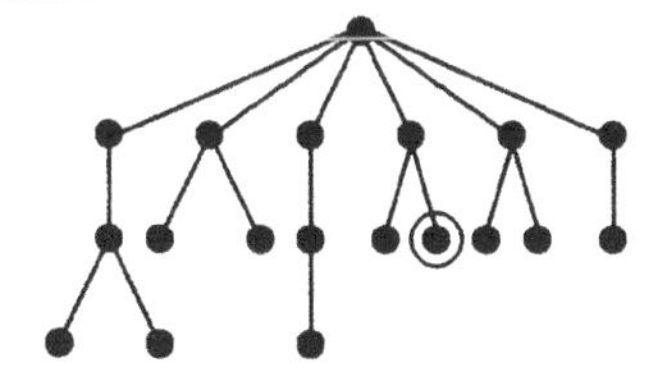

that becomes

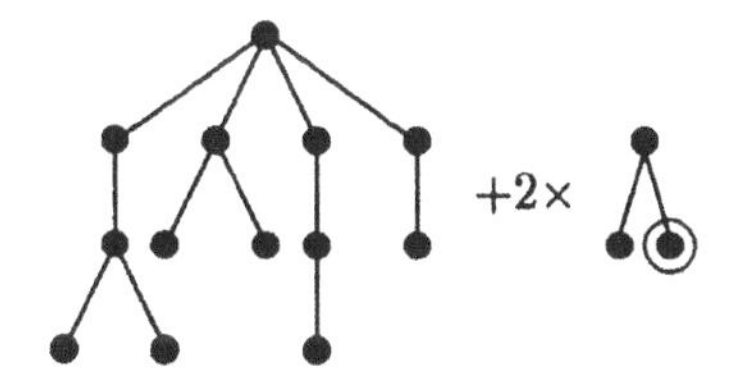

Therefore, we just decompose our arborescence T with a pointed node v into an arborescence T_1 with $n - jd$ nodes and j samples of an arborescence T_2 with d nodes and a pointed node.

But this decomposition defines a 1-1 correspondence between the arborescence T with n nodes and a pointed node that is not the root of T and the triple (T_1, T_2, j) such that

- T_2 is an arborescence with a pointed node,
- T_1 is an arborescence,
- If we denote by d the number of nodes of T_2 then T_1 has $n - jd$ nodes.

Indeed, when we have such a triple (T_1, T_2, j), we obtain the original representation of T with its pointed node v

- by concatenating j samples of T_2 to the root of T_1,
- by representing this arborescence with its j pointed nodes,

- by removing the circle on the $j-1$ rightmost pointed nodes of this representation.

Example :

If we go back to the preceding example

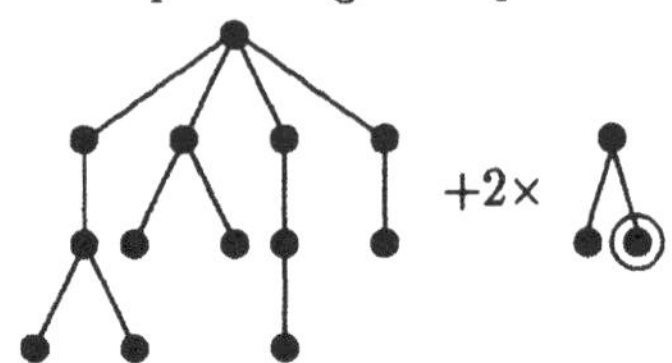

we obtain

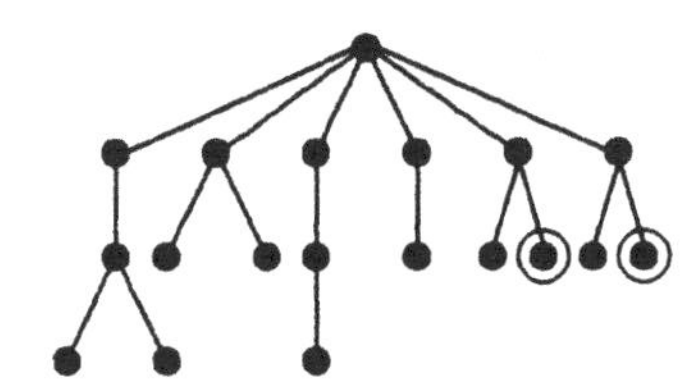

Therefore, in using our coherent representation of arborescences and in removing the rightmost circle :

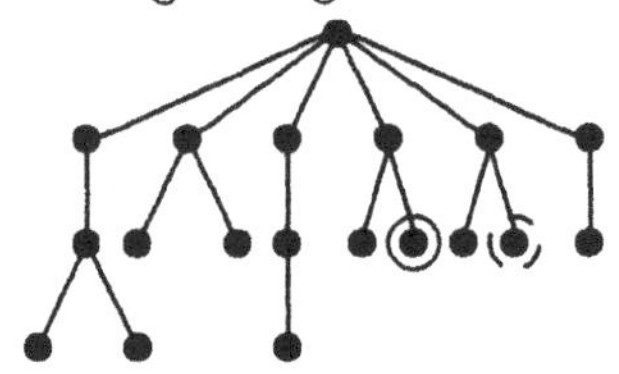

Therefore the number of arborescences with n nodes and a pointed node that is not the root is equal to the number of triples (T_1, T_2, j) which satisfy when we denote by d the number of nodes of T_2 :

- $1 \leq d \leq n-1$,
- $card(T_1) = n - jd$,
- $1 \leq j \leq \lfloor \frac{n-1}{d} \rfloor$ (since $1 \leq card(T_1) \leq n-1$).

So, we obtain when d goes through $[\![1, n-1]\!]$ and j goes through $[\![1, \frac{n-1}{d}]\!]$, the recurrence formula :

$$(n-1)t_n = \sum_{d=1}^{n-1} \sum_{j=1}^{\lfloor \frac{n-1}{d} \rfloor} d\, t_d t_{n-jd}.$$

□

The RANRUT algorithm is a natural continuation of this proof. Indeed, if we want to draw a random arborescence with n nodes it is sufficient :

- to choose a pair of natural integers (j, d) with probability :

$$\frac{1}{(n-1)t_n} d \, t_{n-jd} t_d, \tag{8.1}$$

- to build a random arborescence T_1 with $n - jd$ nodes and a random arborescence T_2 with d nodes,
- to concatenate j samples of the tree T_2 to the root of T_1.

Example :

To generate a random arborescence with four nodes, we obtain :

pair chosen (j, d)	$(1,1)$		$(2,1)$	$(3,1)$	$(1,2)$	$(1,3)$	
with probability :	$\frac{2}{12}$		$\frac{1}{12}$	$\frac{1}{12}$	$\frac{2}{12}$	$\frac{6}{12}$	
generation of T_1 and T_2 with probability	$\frac{1}{2}$	$\frac{1}{2}$	1	1	1	$\frac{1}{2}$	$\frac{1}{2}$
the result arborescence :							
with probability	$\frac{1}{12}$	$\frac{1}{12}$	$\frac{1}{12}$	$\frac{1}{12}$	$\frac{2}{12}$	$\frac{3}{12}$	$\frac{3}{12}$

In the next section we propose an implementation of this algorithm and later we study the complexity of this implementation.

8.3 ALGORITHM

Here, we present the implementation of the RANRUT algorithm that was originally given in [38].

This algorithm can be decomposed into two very different parts.

- First, we compute the values of t_i for each $i \in [\,1, n\,]$. This very expensive step can be done only one time if we want to generate many arborescences with n nodes.
- Then we begin to draw a random arborescence of size n recursively. At each drawing, we choose a pair of integers (j, d), then we generate two arborescences T_1 and T_2 with $n - jd$ and d nodes, finally we concatenate j samples of the tree T_2 to the root of T_1.

The proof that this algorithm randomly and uniformly generates each arborescence with n nodes is the same as the proof of the Property 3.2 (page 135).

8.3.1 The preprocessing step : the computation of the values t_i

Let us define the sequence $(v_1, v_2, \ldots)$ by

$$v_n = \sum_{d|n} d\, t_d.$$

We have now the following proposition.

Proposition 17 *When $n > 1$, we get*

- $v_n = \sum_{d|n} d\, t_d$.
- $t_n = \frac{1}{n-1} \sum_{k=1}^{n-1} t_{n-k} v_k$.

Proof. Indeed, we have

$$\begin{aligned} t_n &= \frac{1}{n-1}\sum_{d=1}^{n-1}\sum_{j=1}^{\lfloor\frac{n-1}{d}\rfloor} d\,t_d t_{n-jd} \\ &= \frac{1}{n-1}\sum_{k=1}^{n-1}\sum_{d\mid k} d\,t_d t_{n-k} \text{ by taking } k = jd \\ &= \frac{1}{n-1}\sum_{k=1}^{n-1} t_{n-k} v_k. \end{aligned}$$

□

Using this Proposition and $t_1 = 1$, we compute iteratively the values of t_i and v_i when i goes from 1 up to n.

8.3.2 The recursive step : the drawing of an arborescence

To draw a random arborescence with n nodes,

- First, we choose a pair of integers (j, d) with the probability :

$$\frac{d}{(n-1)t_n} t_{n-j d} t_d,$$

- Then we generate a random arborescence T_1 with $n - jd$ nodes,
- Next we generate a random arborescence T_2 with d nodes,
- Finally, we concatenate j samples of the tree T_2 to the root of T_1.

The choice of the pair (j, d) can be done as follows:

- first, we generate a random number N in $[\,1, (n-1)t_n]$.
- finally we look for the pair (j, d) which corresponds to N with two nested loops. In the outside loop, d goes from $n - 1$ down to 1, and in the inside loop j goes from 1 up to $\lfloor\frac{n-1}{d}\rfloor$.

Precisely, the function *generate* is defined by

```
N = 1+random ((n − 1)t_n) // draw a randon number in [ 1,(n − 1)t_n ]
d = n
do
   d = d-1
   j = 0
   do
      j = j+1
      N = N-d t_d t_{n-jd}
   while (j < ⌊(n-1)/d⌋) and (N > 0)
while (N > 0)
// the pair (j,d) has just been chosen
T_1 = generate (n − jd)
T_2 = generate (d)
concatenate j samples of T_2 to the root of T_1
return the obtained tree.
```

Remark :

In the outside loop, the number d goes from $n-1$ down to 1. This fact is important since it will determine the average complexity of the algorithm. Indeed the number $d\, t_d t_{n-jd}$ is bigger when d is near $n-1$ than when d is near 1, so it is much more interesting to take d equal to $n-1$ and to decrease its value than to take d equal to 1 and to increase its value.

P. Flajolet et al.[27] have proposed another method which they call the boustrophedion method. This method consists in taking $1, n-1, 2, n-2, \ldots$ as the succesive values of d and generally gives algorithms which are efficient on the average.

8.4 AVERAGE COMPLEXITY ANALYSIS

8.4.1 The preprocessing step

In the preprocessing step of RANRUT the numbers t_n and v_n are computed iteratively using Proposition 17. The corresponding algorithm has two loops : n runs from 1 to n and k or d runs from 1 to n.

It is easily seen that the complexity of this procedure is :

$$\Theta(\sum_{m=1}^{n}\sum_{d=1}^{m-1} 1) = \Theta(1)$$

assuming that the complexity of all classical arithmetic operations is $O(1)$. But, in fact, the multiplication of two numbers with n bits is done in $O(n\, ln(n)$ $Ln_2(n))$ time (see [34])and the global time complexity of the preprocessing step is therefore :

$$O(n^3\, ln(n) Ln_2(n))$$

where $Ln_2(n) = Ln(Ln(n))$ and $Ln(n) = max(1, Log(x))$. All the numbers t_i, $(1 \leq i \leq n)$ have to remain stored. The storage cost is clearly $O(n)$ in case 1 and $O(n^2)$ in case 2.

8.4.2 Complexity of the random drawing of a pair (j, d)

As mentioned in Section 2, a pair (j, d) is chosen with probability $prob_n(j, d)$ given by Formula (8.1). RANRUT has two loops, in the first one d runs from $n-1$ down to 1 and in the second j runs from 1 up to $\lfloor \frac{n-1}{d} \rfloor$. Now we generate a tree T_1 on $(n - jd)$ vertices and a tree T_2 on d vertices. We make j copies of T_2 as mentioned in Section 2.

Remark :

> It is clear that the worst case complexity is reached if the pair $(1, n)$ is the last one chosen. This gives a worst case complexity in $O(n\, ln(n))$ to choose a pair (j, d) in case 1 and $O(n^2\, ln^2(n) Ln_2(n))$ in case 2.

Behavior of the probabilities

Using (8.1) and the Property 3.4, we obtain :

$$prob_n(j, d) = \Theta(\frac{\sqrt{n}}{\sqrt{d}(n - jd)^{\frac{3}{2}} \rho^{(j-1)d}})$$

when $j\,d < n$. In addition, this shows that :

$$\sum_{j=1}^{\lfloor \frac{n-1}{d} \rfloor} prob_n(j,d) \quad = \quad \Theta(\frac{\sqrt{n}}{\sqrt{d}(n-d)^{\frac{3}{2}}}) = \Theta(prob_n(1,d)). \qquad (8.2)$$

In fact, in using the convention that $\sum_{j=a}^{b} f(j) = 0$ when $a > b$ whatever the values $f(j)$ we get :

$$\sqrt{\frac{d}{n}} \sum_{j=1}^{\lfloor \frac{n-1}{d} \rfloor} prob_n(j,d) \sim \sum_{j=1}^{\lfloor \frac{n-1}{d} \rfloor} \frac{1}{(n-jd)^{\frac{3}{2}} \rho^{(j-1)d}})$$

$$= \frac{1}{(n-d)^{\frac{3}{2}}} + \sum_{j=2}^{\lfloor \frac{n-1}{2d} \rfloor} \frac{\sqrt{n}}{\sqrt{d}(n-jd)^{\frac{3}{2}} \rho^{(j-1)d}} + \sum_{j=\lfloor \frac{n-1}{2d} \rfloor +1}^{\lfloor \frac{n-1}{d} \rfloor} \frac{\sqrt{n}}{\sqrt{d}(n-jd)^{\frac{3}{2}} \rho^{(j-1)d}}$$

$$= \frac{1}{(n-d)^{\frac{3}{2}}} + O\left(\frac{1}{n^{\frac{3}{2}}} \sum_{j=2}^{\lfloor \frac{n-1}{2d} \rfloor} \frac{1}{\rho^{(j-1)d}}\right) + O\left(\frac{1}{\rho^{d \lfloor \frac{n-1}{2d} \rfloor}} \sum_{j=\lfloor \frac{n-1}{2d} \rfloor +1}^{\lfloor \frac{n-1}{d} \rfloor} 1\right)$$

$$= \Theta\left(\frac{1}{(n-d)^{\frac{3}{2}}}\right).$$

Case 1 : the complexity of the arithmetic operations is $O(1)$

We can now study the average complexity that we need to choose a pair (j,d).

Theorem 30 *The average complexity of RANRUT to draw a pair (j,d) is in $\Theta(\sqrt{n})$.*

Proof. Let $C_n(j,d)$ be the time necessary to choose a pair (j,d) :

$$C_n(j,d) = \sum_{\overline{d}=d+1}^{n-1} \left\lfloor \frac{n-1}{\overline{d}} \right\rfloor + j$$

since for each $\overline{d}$ we have $\left\lfloor \frac{n-1}{\overline{d}} \right\rfloor$ possibilities for j, except when $\overline{d} = d$ where we do j tests.

Therefore, using $j \leq \frac{n-1}{d} \leq n-d$ we have

$$\begin{aligned} C_n(j,d) &= \sum_{\bar{d}=d+1}^{n-1} \frac{n-1}{\bar{d}} + O(n-d) \\ &= \int_d^{n-1} \frac{n-1}{x} dx + O(n-d). \end{aligned}$$

$$C_n(j,d) = \Theta(n\, ln(\frac{n}{d})). \tag{8.3}$$

Let $\overline{C_n}$ be the average time necessary for the random generation of the pair (j,d) :

$$\overline{C_n} = \sum_{d=1}^{n-1} \sum_{j=1}^{\lfloor \frac{n-1}{d} \rfloor} prob(j,d) C_n(j,d).$$

Using (8.2) and (8.3) :

$$\begin{aligned} \overline{C_n} &= \Theta\left(\sum_{d=1}^{n-1} prob(1,d) n\, ln(\frac{n}{d})\right) \\ &= \Theta\left(\sum_{d=1}^{n-1} \frac{n^{\frac{3}{2}}\, ln(\frac{n}{d})}{\sqrt{d}(n-d)^{\frac{3}{2}}}\right) \\ &= \Theta\left(\sum_{d=1}^{\lfloor \frac{n-1}{2} \rfloor} \frac{ln(\frac{n}{d})}{\sqrt{d}} + \sum_{d=\lfloor \frac{n-1}{2} \rfloor + 1}^{n-1} \frac{1}{(n-d)^{\frac{1}{2}}}\right) \\ &= \Theta\left(\int_1^{\frac{n}{2}} \frac{ln(n) - ln(x)}{\sqrt{x}} dx + \sqrt{n}\right) \\ &= \Theta([2\, ln(n)\sqrt{x} - 2\, ln(x)\sqrt{x} + 4\sqrt{x}]_1^{\frac{n}{2}} + \sqrt{n}) \\ &= \Theta(\sqrt{n}). \end{aligned}$$

□

Remark : If we let d go from 1 up to $n-1$ in the main loop of RANRUT, we find a value $\overline{C_n}$ in $\Theta(n\, ln(n))$. This is not a good strategy.

Case 2

Each acceptance or rejection step of a pair (j,d) consists of one addition of a number with (approximately) n bits and one multiplication of two numbers the first with $(n-d)$ bits, the second with approximatively d bits.

We get, using similar methods,

Theorem 31 *The average complexity to choose a pair* (j,d) *is in* $\Theta(n^{\frac{3}{2}} ln(n) Ln_2(n))$

8.4.3 Choosing the size of the subtrees of a vertex : complexity analysis

Assume that we have to choose the size of the subtrees of a vertex of size m. We have to apply RANRUT to the root of these subtrees N times, where N is bounded by the final number of sons of this vertex. The size of the subtrees on which we apply the algorithm is, of course, decreasing but we assume that the algorithm is always applied to a tree of size m. This provides an upper bound for the complexity.
The average complexity is bounded above by :

$$\frac{1}{t_m} \sum_{P \text{ trees with } m \text{ vertices}} \overline{C_m} R(P) \tag{8.4}$$

where $R(P)$ is the number of sons of the tree P.
(8.4) is in fact the average number of sons of a tree of size m times $\overline{C_m}$. Meir and Moon [37] have proved that this number tends to a constant K $(= 2.71\ldots)$ as m goes to infinity.
Therefore, we have proved that the average complexity for choosing the size of the subtrees of a node of size m is $\Theta(\overline{C_m})$. Indeed we know that we will have to choose at least a pair (j,d), and this requires an average complexity $\overline{C_m}$.

8.4.4 The global complexity of a drawing

We give here an upper bound for the average complexity of each of the cases studied previously. For this purpose, we assume that for each vertex the drawing has been done even if the vertex is built via a simple copy of another part of the tree. Therefore the average complexity is bounded above by

$$\frac{K}{t_n} \sum_{\substack{\text{T trees} \\ \text{with n vertices}}} \sum_{\substack{\text{w vertex} \\ \text{of the tree T}}} \overline{C_{|w|}} \qquad (8.5)$$

where $|w|$ is the number of vertices in the subtree of T defined by w and $\overline{C_m}$ is the average time spent at the top of a tree of size m for choosing the right values of the number of vertices found in the subtrees.
First, we examine the distribution of the vertices with a given number of edges in their subtrees.
Let $N_n(m)$ be the number of vertices corresponding to a subtree of size m, vertices which we can find in all trees of size n. It is easy to see that if we replace this subtree by a simple vertex, we obtain a tree of size $n - m + 1$. Therefore $N_n(m) - N_{n-m+1}(m)$ is bounded above by $g_{n-m+1}t_m$ where g_{n-m+1} is the number of leaves in the trees of size $n - m + 1$ (identical leaves are counted once). Indeed, when we choose a subtree S of size m in a tree of size m and when we cut it off, we obtain a tree U of size $n - m + 1$. If we can find many representatives of the subtree S, the other representatives of S are always in the tree U. Therefore the number of subtrees of size m, which are in all trees of size m, is bounded above by the sum of the number of subtrees of size m which are in the trees of size $n - m + 1$ and of $g_{n-m+1}t_m$.

Behavior of g_n

The function g_n defined above has the following properties :

- $g_0 = 0; g_1 = 1$
- let g (resp. y) be the ordinary generating function of g_n (resp.t_n, the number of trees of size n, $t_0 = 0$), we have :

$$g(x) \quad = \quad x + g(x)\,y(x) \qquad (8.6)$$

A leaf is the root of a tree of size 1 or is in one of the subtrees of the root. In the latter case, we split this tree in two, the subtree which contains this leaf and another tree.

Therefore :

$$g(x) = \frac{x}{1 - y(x)}$$

Using Meir and Moon's [37] result, we find that :

$$g_n = \Theta(\frac{\rho^n}{\sqrt{n}})$$

and :

$$N_n(m) - N_{n+1-m}(m) = \Theta(\frac{\rho^n}{m^{\frac{3}{2}}\sqrt{n+1-m}})$$

finally we get :

$$N_n(m) = \Theta(\frac{\rho^n}{m^{\frac{3}{2}}\sqrt{n+1-m}})$$

Complexity of RANRUT in case 1

Theorem 32 *The average complexity of RANRUT is in $O(n\,ln(n))$.*

Proof. The calculations are close to those made in [37] for the average height of a vertex. Using (8.5), we obtain an upper bound for the average complexity $\overline{C_n}$:

$$\overline{C_n} = O\left(\frac{1}{t_n}\sum_{m=1}^{n} N_n(m)\overline{C_m}\right).$$

Therefore, since $\overline{C_m} = \Theta(\sqrt{m})$:

$$\begin{aligned}
\overline{C_n} &= O(n^{\frac{3}{2}} \sum_{m=1}^{n} \frac{1}{m\sqrt{n-m}}) \\
&= O(n \sum_{m=1}^{\lfloor \frac{n-1}{2} \rfloor} \frac{1}{m}) + O(n^{\frac{1}{2}} \sum_{\lfloor \frac{n+1}{2} \rfloor}^{n-1} \frac{1}{\sqrt{n-m}}) \\
&= O(n\,lnn).
\end{aligned}$$

□

Remark :
The complexity of the preprocessing step is $O(n^2 ln(n))$ which is, in this case, the main part of the total complexity.

Complexity of RANRUT in case 2

Theorem 33 *The average complexity of RANRUT is upper bounded by* $O(n^2\, ln(n)\, Ln_2(n))$.

Proof. The calculations are the same as above and the details are omitted:

$$\begin{aligned}
\overline{C_n} &= O(n^{\frac{3}{2}} \sum_{m=1}^{n-1} \frac{ln(m)Ln_2(m)}{\sqrt{n-m}}) \\
&= O(n \sum_{m=1}^{\lfloor \frac{n-1}{2} \rfloor} ln(m)Ln_2(m) + n^{\frac{3}{2}}\, ln(n)Ln_2(n) \sum_{m=\lfloor \frac{n+1}{2} \rfloor}^{n-1} \frac{m}{\sqrt{n-m}}) \\
&= O(n^2\, ln(n)Ln_2(n)).
\end{aligned}$$

□

Remark about the copying of trees
Instead of copying trees, we made a pseudo-redrawing of them. It is a question whether the complexity is increased significantly. The answer is no.
Indeed, we have previously seen that the average complexity can be bounded above by :

$$O(\frac{1}{t_n} \sum_{m=1}^{n-1} N_n(m)\overline{C_m}).$$

Let $F(u,x)$ be the generating function of $U_n(m)$ the number of nodes with size m belonging to trees of size n (identical nodes are counted once) .

$$F(u,x) = \sum_{n=1}^{\infty} \sum_{m=1}^{\infty} U_n(m) x^n u^m.$$

Then we have :

$$F(u,x) = y(ux) + F(u,x)y(x).$$

This gives us :

$$F(u,x) = \frac{y(ux)}{1-y(x)}$$

with the series expansion :

$$\frac{1}{1-y(x)} = \sum_{n=1}^{\infty} y_n x^n$$

we get :

$$U_n(m) = t_n y_{n-m}.$$

Using the same method as for g_n, we obtain :

$$U_n(m) = \Theta(\frac{\rho^n}{m^{\frac{3}{2}}\sqrt{n-m}}) = \Theta(N_n(m)).$$

This gives us a lower bound for the average complexity. Indeed we know that we have to choose (j,d) for all the nodes which are counted by $N_n(m)$.
And therefore we have a lower bound for the average complexity :

$$\Omega(\frac{1}{t_n}\sum_{m=1}^{n-1} U_n(m)\overline{C_m}). \tag{8.7}$$

Using the upper bounds (8.5) and (8.7), the average complexity is then :

$$\Theta(\frac{1}{t_n}\sum_{m=1}^{n-1} N_n(m)\overline{C_m}).$$

8.5 ANOTHER EXAMPLE : THE GENERATION OF A BINARY ARBORESCENCE

In this section, we will see that a binary arborescence (ie. an arborescence whose nodes have zero or two children) can be generated using the same method.

Let us denote by t'_n the number of binary arborescences with n inner nodes, then

Theorem 34 *We have*

- $t'_0 = 1$,
- $\forall n \in \mathbb{N}^\star, t'_{2n} = \sum_{i=1}^{n-1} t'_i t'_{2n-i-1}$,
- $\forall n \in \mathbb{N}, t'_{2n+1} = \sum_{i=1}^{n-1} t'_i t'_{2n-i} + \frac{1}{2}({t'_n}^2 + t'_n)$,
- *and* $t_n \sim c\frac{\rho^n}{n^{\frac{3}{2}}}$ *with* $\rho = \sqrt[24]{2090918}$ *and* $c = 2.3532\ldots$.

Proof. The property 1 of this theorem is straightforward.

A binary arborescence with $2n$ inner nodes has a root that has one subtree child that has strictly less than n inner nodes : i, and that has another subtree child with $2n - i$ inner nodes. This shows property 2 of this theorem.

To prove property 3 of this theorem, we remark that a binary arborescence with $2n + 1$ inner nodes is either an arborescence whose root has a subtree child that has strictly less than n inner nodes (there exist $\sum_{i=1}^{n-1} t_i t_{2n-i}$ such arborescences) or an arborescence whose root has two subtrees with n inner nodes. Now we need to count the number of the latter arborescences : $\frac{1}{2}t_n^2$ counts once time the arborescences that have a root with two different subtrees and $\frac{1}{2}$ times the arborescences having the same root subtrees; therefore this number is equal to $\frac{1}{2}t_n^2 + \frac{1}{2}t_n$.

The asymptotics of t'_n has been studied in [39]; this proof uses complex analysis and is too technical to be presented in this book. □

The algorithm of generation is now straightforward. First we compute the values of the t_i when $i \in [\![1, n]\!]$; the complexity of this preprocessing step is in $O(n^2)$.

Then we apply the following recursive procedure :

N = 1+random $(2t_n)$ *// random number in* $[\![1, 2t_n]\!]$
d = n
do
  d = d-1
  N = N-2 $t_d t_{n-1-d}$
while (N > 0) and $(d \geq \lceil \frac{n}{2} \rceil + 1)$

```
if N > 0
    // this case appears when n = 2n' + 1 and when the two root subtrees
    // has the same number of internal nodes
    mil = ⌊n/2⌋
    if N > t_mil
        d = mil
    else
        T1 = generate (mil)
        return the arborescence whose root has T1 and T1 as children
T1 = generate (n - 1 - d)
T2 = generate (d)
return the arborescence whose root has T1 and T2 as children.
```

The study of the complexity of this algorithm can be done with the same methods used for random arborescences. We obtain the following result :

Theorem 35 *The recursive algorithm that builds a binary random arborescence has the following average complexities :*

- *case 1 : all arithmetic operations (addition, multiplication etc...) have a complexity in* $O(1)$
    1. *preprocessing : time complexity* $\Theta(n^2)$ *and space complexity* $\Theta(n)$.
    2. *recursive part : time complexity* $\Theta(n\, ln(n))$.
- *case 2 : the arithmetic operations have a time complexity in* $O(n\, ln(n)\, Ln_2(n))$,
    1. *preprocessing : time complexity* $\Theta(n^3\, ln(n) Ln_2(n))$, *space complexity* $\Theta(n^2)$.
    2. *recursive part :* $\Theta(n^2\, ln(n) Ln_2(n))$.

8.6 CONCLUSION

In this chapter we have presented a recursive algorithm that builds a random arborescence and has the following average complexities :

- case 1 : all arithmetic operations (addition, multiplication etc...) have a complexity in $O(1)$

1. preprocessing : time complexity $\Theta(n^2)$ and space complexity $\Theta(n)$.
2. recursive part : time complexity $\Theta(n\, ln(n))$

- case 2 : the arithmetic operations have a time complexity in. $O(n\, ln(n)\, Ln_2(n))$ ([34]),

  1. preprocessing : time complexity $\Theta(n^3\, ln(n) Ln_2(n))$, space complexity $\Theta(n^2)$.
  2. recursive part : $\Theta(n^2\, ln(n) Ln_2(n))$.

The same methods can be applied to generate a random binary arborescence with the same complexities.

In fact, Flajolet et al. have proved in [27] that this method can be applied to build a tree with n nodes whenever the family of trees we want to build can be described by a grammar of trees. Their algorithm has a preprocessing step with worst case complexity in $O(n^2)$ and a recursive step whose average complexity is in $O(n\, ln(n))$ in case 1.

For case 2, as we will see in the next chapter, it would be interesting to compute the values t_n with low accuracy and to begin the recursive part. Sometimes with a low probability, we would need to stop the generation because we only know the values t_n with too low accuracy, in which case we compute exactly the values t_n and we complete the drawing.

9

GENERATION OF TREES WITH A GIVEN HEIGHT AND SOME TRICKS ABOUT COMPLEXITY

Abstract

We present in this chapter an algorithm for drawing a random binary tree of a given height. This is a classical recursive algorithm such as those presented in the previous chapter. But, this time, we study how the complexity of such algorithms can be decreased by using numbers with low accuracy. We show that the same methods can be used to decrease the complexity of the generation of an arborescence.

9.1 INTRODUCTION

In this chapter, we study the problem of the generation of binary tree of height h. First, we present some combinatorial results about these trees, then we propose a recursive algorithm to generate such a tree. Finally, we show how to improve its complexity by using some numbers with low accuracy.

In fact, this chapter exploits an idea of P. Zimmermann : we can often do the computations using numbers with low accuracy and when we detect that a problem can be caused by the low accuracy numbers, we use the exact numbers. If a problem appears with a small probability, we obtain an algorithm that can have an average complexity that is much better that the algorithm that always uses the exact numbers.

In the last section, we show how this method can be applied in order to decrease the complexity of the generation of an arborescence.

9.2 GENERATION OF A BINARY TREE OF HEIGHT H

First, we give some basic results about the binary trees of height h, then we study how to generate them by using a recursive method and how to improve the average complexity.

9.2.1 Definitions and basic notions

Let

- $T(h)$ be the number of binary trees of height h,
- $C(h)$ the number of binary trees of height less or equal to h,
- $N(h)$ the number of nodes in the set of all binary trees of height h,
- $M(h)$ the number of nodes in the set of binary trees of height less than or equal to h.

Proposition 18 *We get:*

- $T(1) = 1, C(1) = 1, N(1) = 1, M(1) = 1,$
- $C(h+1) = C(h) + T(h+1),$
- $C(h+1) = 1 + C(h)^2,$
- $C(h) = \alpha^{2^h} + O\left(\frac{1}{\alpha^{2^{h+1}}}\right)$ *with* $\alpha \in \boldsymbol{R}, \alpha = 1.22590244\ldots,$
- $M(h+1) = M(h) + N(h+1),$
- $M(h+1) = 2C(h)M(h) + C(h+1),$
- $M(h) = \gamma 2^h C(h) - 1 + O\left(\frac{2^h}{\alpha^{2^{h+1}}}\right)$ *with* $\gamma \in \boldsymbol{R}.$

Proof. The relations for computing recursively the values of $C(h)$ and $T(h)$ are very simple. They are based on the following principles:

- a binary tree of height less than $h+1$ is either a leaf or a root with two children of height less than h,

- the nodes in the trees of height less than $h+1$ are either at the root (the term $C(h+1)$), or in the left (or right) subtree of the root.

The asymptotic behavior of $C(h)$ and $M(h)$ is derived from the study of the recurrence relations $C(h+1) = 1+C(h)^2$ and $M(h+1) = 2C(h)M(h)+C(h+1)$.
□

We will also need some practical results in our algorithm. These results are easy to prove by induction:

Proposition 19 *Let $\beta = \sqrt[8]{2}$, we have the following relations:*

- $\forall h \geq 2, C(h) \geq \beta^{2^h},$
- $\forall h \geq 3, \dfrac{1}{T(h)} \leq \dfrac{1}{\beta^{2^h}},$
- $\forall h \geq 3, \dfrac{C(h-1)}{C(h)} \leq \dfrac{1}{\beta^{2^{h-1}}}.$

9.2.2 Generation using recursivity

The obvious method to generate a random binary tree of height h is :

- First to compute the values of $T(i)$ when $i \in [\![1, h]\!]$.
- Then to define a recursive function which builds a random binary tree of height i. This function must consider the fact that there are $2i-3$ ways to choose the heights of the two children subtrees of the root. These $2i-3$ possibilities and the number of trees which correspond to each possibility are represented in the following array :

  A binary tree of height i is

number of trees	type of trees	
$T(i-1)^2$	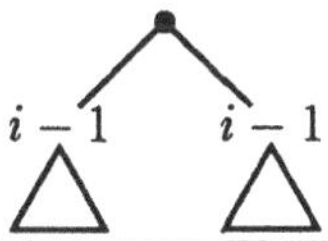 $i-1$ $i-1$	
$T(i-1).T(i-2)$	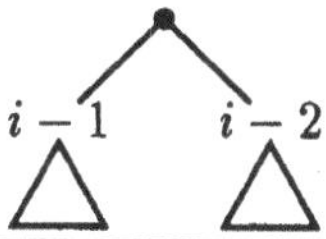 $i-1$ $i-2$	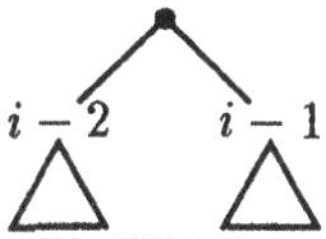 $i-2$ $i-1$
$T(i-1).T(i-3)$	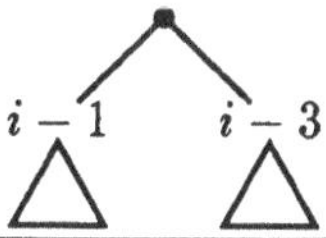 $i-1$ $i-3$	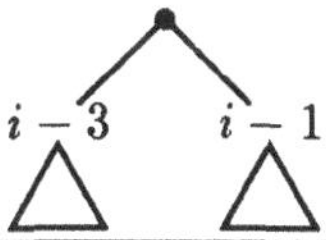 $i-3$ $i-1$
...	...	...
$T(i-1).T(1)$	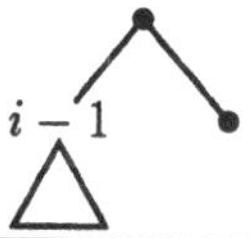 $i-1$	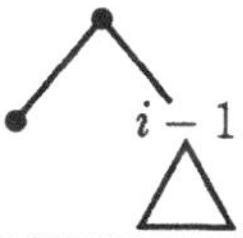 $i-1$

So, if we want to write this algorithm in a more rigorous form, we have an algorithm composed of two different steps defined by

- A preprocessing step of computing the values $C(i)$ for all i in $[1, h]$ by using the formulas $C(1) = 1$ and $C(i+1) = C(i)^2 + 1$.
- A recursive function that builds a tree of height i and which is defined as follows:
  - First draw a number N in $[1, C(i-1) + C(i-2)]$,
  - Then choose the height l of the left subtree of the root and the height r of the right subtree of the root by using the following array :

	choose
$If\, 2C(i-2) < N$	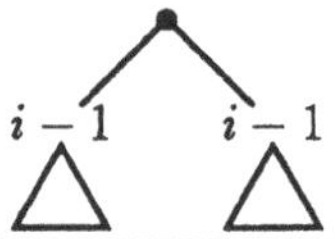 $i-1$ $i-1$
$If\, C(i-2)+C(i-3) < N \leq 2C(i-2)$	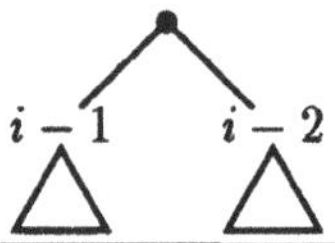 $i-1$ $i-2$
...	...
$If\, C(1) < N \leq 2C(1)$	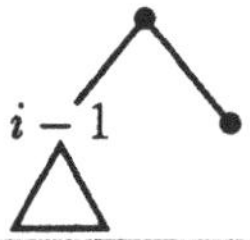 $i-1$
$If\, N \leq C(1)$	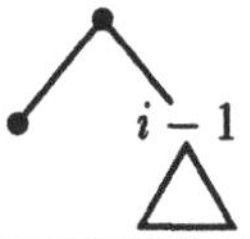 $i-1$

- draw a tree T_l of height l and a tree T_r of height r.
- return the tree formed by a root whose children are T_l and T_r.

Remark :

The values of this array are obtained by dividing the values of the penultimate array by $T(i-1)$ and remarking that $C(1) = T(1)$, $2C(1) = T(1) + T(1)$, $C(1) + C(2) = T(1) + T(1) + T(2)$, ...

We can now use the fact that the asymptotic behavior of $C(h)$ is in α^{2^h} to remark that

- the complexity of the preprocessing step is in $\Omega(\alpha^h\, h\, ln(h))$ when we use the real complexities of the basic operations,
- for a binary tree of height i, the complexity of choosing the heights of the root's children is in $\Omega(2^i)$ since we need to draw a number that has approximately $ln(\alpha)2^i$ bits.

We will show in the next subsection how to improve these two complexities. This turns out to be rather easy to do.

9.2.3 Another algorithm of generation

In this subsection we show how to improve the complexity of the preceding algorithm. In fact, we show that the drawing can begin without doing a preprocessing step and that the computations of the exact values of the $C(i)$'s must be done with low probabilities.

In order to do this, first we prove the following proposition :

Proposition 20 *When we draw a binary tree of height i, the probability that we choose a height $i-1$ for the two children of the root is in*

$$1 - \frac{2}{\alpha^{2^{i-2}}} + O(\frac{1}{\alpha^{2^{i-1}}}).$$

Proof. Indeed, if we compute the probability of choosing two trees of height $i-1$, we find a probability :

$$\begin{aligned} \frac{T(i-1)^2}{T(i)} &= \frac{C(i-1)-C(i-2)}{C(i-1)+C(i-2)} \\ &= 1 - \frac{2C(i-2)}{C(i-1)+C(i-2)} \\ &= 1 - \frac{2}{\alpha^{2^{i-2}}} + O(\frac{1}{\alpha^{2^{i-1}}}). \end{aligned}$$

□

This proposition is very important because it claims that when we draw a binary tree of height i, the choice of the heights $i-1$ for the two children of the root is done with high probability. But this formula is not sufficient because it furnishes only an estimation of the asymptotic behavior for this probability. The following proposition gives a lower bound of this probability.

Proposition 21 *When we draw a binary tree of height i, the probability that we choose the height $i-1$ for the two children of the root is greater than*

$$1 - \frac{2}{\beta^{2^{i-2}}}$$

when $i > 3$.

Proof. We use Proposition 19 to obtain a lower bound of $\frac{2C(i-2)}{C(i-1)+C(i-2)}$. □

This means that the probability of a different choice is less than $\frac{2}{\beta^{2^{i-2}}}$.

We use this proposition to obtain a procedure to choose the pair (r, l) which will give the height of the left and the right subtree of the root when we want to generate a tree of height i (with $i > 3$). This algorithm consists of:

- drawing a number N between 0 and 1 bit after bit,
- then multiplying this number by $C(i-1) + C(i-2)$
- finding the pair (j, d) that corresponds to this number.

We stop as soon as we can, for instance when N is greater than $\frac{2}{\beta^{2^{i-2}}}$, this will mean that the pair (r, l) is indeed the pair $(i-1, i-1)$.

This gives us a lazy drawing to find a pair (r, l) when $i > 3$:

- Begin to draw the i first bits of N, choose the pair $(i-1, i-1)$ as soon as you draw a 1 bit and stop the procedure.
- Compute the value $\lfloor ln(\beta).2^i \rfloor$, and begin to draw the next bits until you draw a 1 bit or have drawn $\lfloor ln(\beta).2^i \rfloor$ 0 bits. In the first case, return the pair $(i-1, i-1)$.
- Compute the values of the $C(j)$'s for $j \in [1, i-1]$.
- Continue the drawing of a number N until you have drawn as many bits as the number $C(i-1) + C(i-2)$ has, multiply N by $C(i-1) + C(i-2)$, look for the choice of a pair (r, l). If you can, return the chosen pair (r, l).
- Continue the drawing of N until N has two times the number of bits of $C(i-1) + C(i-2)$, multiply N by $C(i-1) + C(i-2)$, ... If you can choose a pair (r, l), return the chosen pair.
- ....

- Continue the drawing of N until N has n times the numbers of bits of $C(i-1)+C(i-2)$, multiply N by $C(i-1)+C(i-2)$, ... If you can choose a pair (r,l), return the chosen pair.
- ....

Of course, this algorithm has infinite worst case complexity. But for its average complexity, we get the following result:

Theorem 36 *For a binary tree of height i, the choice of the heights of the root's children has an average complexity in $O(ln(i))$.*

Proof. We study the complexity of each step :

- In the first step of this algorithm, we need to compare the value of a counter at each step, this accounts for a complexity in $O(ln(i))$, then we draw a first bit with probability 1, a second bit with probability $\frac{1}{2^1}$, ..., a ith bit with probability $\frac{1}{2^{i-1}}$.

  Therefore, this gives us an average complexity in

$$ln(i) + \frac{ln(i)}{2} + \frac{ln(i)}{2^2} + \ldots + \frac{ln(i)}{2^{i-1}} = O(ln(i)).$$

- The second step is only executed with a probability $\frac{1}{2^i}$ which is the probability that we have drawn i 0 bits in the first step.

  In this step, we must compute the value of $\lfloor ln(\beta)2^i \rfloor$, this accounts for a complexity equal to $i\,ln(i)\,Ln_2(i)$. Finally, we begin drawing the next bits of N and compare a counter to the value $\lfloor ln(\beta)2^i \rfloor$.

  Therefore, we obtain an average complexity of this step in

$$\frac{i\,ln(i)\,Ln_2(i) + i + \frac{i}{2} + \frac{i}{2^2} + \ldots}{2^i} = O(1).$$

- The third step is executed with a probability $\frac{2}{\beta^{2^i}}$.

  In this step we compute the values of $C(j)$ for each $j \in [\![1, i-1]\!]$.

  This gives us a complexity in

$$\frac{2\,2^i\,i\,ln(i)}{\beta^{2^i}} = O(1).$$

- The fourth step is executed with the same probability as the third one, this is $\frac{2}{\beta^{2^i}}$.

  After each bit drawing, we need to compare a counter with a value that has $O(2^i)$ bits. Finally, we multiply the number N by $C(i-1)+C(i-2)$ this accounts for a complexity in $O(2^i\, i\, ln(i))$ and we compare this number with at most $2i-3$ values with $O(2^i)$ bits.

  Therefore the average complexity of this step is in

$$\frac{2.2^i + 2^i\, i\, ln(i) + (2i-3)2^i}{\beta^{2^i}} = O(1).$$

- The fifth step is executed with probability $O(\frac{2i}{\alpha^{2^i}})$, this gives us an average complexity in

$$2i\frac{2^{i+1} + 2^{i+1}\,(i+1)\,ln(i+1) + (2i-3)2^{i+1}}{\alpha^{2^i}}.$$

- ...,
- The seventh step is executed with probability $O(\frac{2i}{\alpha^{(n-1)2^i}})$, this gives us an average complexity in

$$O(2i\frac{(n-1)2^{i+1} + (n-1)2^{i+1}\,(i+1)\,ln(i+1) + (n-1)(2i-3)2^{i+1}}{\alpha^{(n-1)2^i}}).$$

- ....

When we sum all these complexities, we find that the average complexity of the procedure is in $O(ln(i))$ as claimed. □

Therefore, if we compare this algorithm with the first version, we no longer have a preprocessing step. And we have obtained a procedure that finds a pair (l, r) for the children of a tree of height i in average time $O(ln(i))$ instead of $\Omega(2^i)$.

Of course, the same kind of methods can be used to generate a k-ary tree: we need in this case to choose the height of each child of the root from the left to the right. In this case we can also use the fact that the probability that the height of all children of the root of a tree of height i is $i-1$ is very great.

Arborescences of height h can also be generated in the same way since it can be shown that the algorithm needs to copy two trees with very low probability.

9.3 GENERATION OF AN ARBORESCENCE

We present in this section a method to decrease the average complexity of the RANRUT algorithm which we have studied in the preceding section. This method found by P. Zimmermann computes the coefficients t_n with low accuracy when we begin the drawing. Sometimes, when we can not be sure that the drawing is correct, we stop, we compute the real values of the t_n and we continue the drawing.

9.3.1 Computation of the values of t_i

We have seen in the preceding chapter that the values t_i can be computed using the two following recurrence formulas :

$$v_i = \sum_{d|i} d\,t_d,$$

$$t_i = \frac{1}{i-1}\sum_{k=1}^{i-1} t_{i-k} v_k$$

and $t_i = 1$.

We want now to compute all these numbers with a low accuracy. We represent each approximative number in the form $b2^m$ where b is a real in $[0.5, 1]$ and m is an integer.

In our case, t_i and v_i will be stored using numbers with b having $\lceil 5ln(n)\rceil + 1$ bits. Each computation will be done using numbers having $\lceil 10ln(n)\rceil$ bits in order to get a precision of $2n\frac{1}{n^{10}}$. Then we keep the most important $\lceil 5ln(n)\rceil+1$ bits of the result.

This means that each time we do a computation, instead of finding a number N, we find a number which is between $N(1-\frac{1}{n^5})$ and N.

Now denote by $\underline{t_i}$ and $\underline{v_i}$ approximate values of the coefficients t_i and v_i that we have computed. We will try to find some values ϵ_i et γ_i such that

$$v_i(1-\epsilon_i) \leq \underline{v_i} \leq v_i$$

$$\text{and } t_i(1-\gamma_i) \leq \underline{t_i} \leq t_i.$$

We can take of course $\epsilon_1 = \gamma_1 \geq 0$ and we will define recursively the other coefficients. For instance, suppose that we have defined some values for the coefficients $\gamma_1, \gamma_2, \ldots, \gamma_i$, we have then

$$v_i(1-\frac{1}{n^5})\min_{d=1\,to\,i}(1-\gamma_d) \leq (1-\frac{1}{n^5})\sum_{d|i} d\,t_d(1-\gamma_d) \leq (1-\frac{1}{n^5})\sum_{d|i} d\,\underline{t_d}$$

$$\leq \underline{v_i} \leq \sum_{d|i} d\,\underline{t_d} \leq v_i.$$

Therefore, we can take when $i \geq 1$

$$\epsilon_i \geq \max_{d=1\,to\,i} \gamma_d + \frac{1}{n^5}.$$

In the same way, we can show that we can take

$$\gamma_i \geq \max_{k=1\,to\,i-1}(\epsilon_k + \gamma_{i-k}) + \frac{1}{n^5}.$$

Proposition 22 *We can take*

$$\epsilon_i = \frac{2i-1}{n^5},$$
$$\gamma_i = \frac{2i-2}{n^5}.$$

Proof. Indeed, we only need to show that these values satisfy the two preceding inequalities. This is straightforward. □

This proposition allows us to show that we compute the approximated values t_i such that

$$t_i(1-\frac{2}{n^4}) \leq \underline{t_i} \leq t_i$$

for each i in $[\,1, n\,]$.

We see in the next section how to use these values to draw a number.

9.3.2 The choice of a pair (j, d)

In order to choose a pair (j, d) which corresponds to a number n we draw a random number N in the interval $[\,1, (n-1)t_n\,]$ then we look in which interval this number N appears. This gives us a pair (j, d).

In this case, the situation is more intricate because we only know the values of t_n with low accuracy.

- First, we begin to draw the $\lfloor ln(\frac{n^4}{2}) \rfloor$ most important bits of a number between 1 and $(n-1)t_n$, if we are not sure that this number is smaller than or equal to $(n-1)t_n$, we go to the third step.
- Then, we look if we can prove that the number N is in a known interval. This can not always be done since we have computed the values of the t_i's with low accuracy. If an interval can be found, we return the corresponding pair (j, d). In the other case we go to Step 3.
- We compute the values of the coefficients t_i exactly. Then we finish drawing our number N and we find in which interval it remains. We return the corresponding pair (j, d).

Let p be the number $\lfloor ln(\frac{n^4}{2}) \rfloor$. We give now a detailed version of the algorithm to choose a pair (j, d):

step 1: We generate a number N between 0 and $(n-1)t_n - 1$ by drawing its p most important bits. We draw this number one bit after the other and we stop either when we have generated p bits or when we are sure that the number that we are generating will be greater than $(n-1)t_n - 1$.

Three cases can arise :

- We have generated a number which is greater than or equal to $(n-1)t_n$, we go back to the beginning.
- We do not already know if the number that we are generating will be greater than or equal to $(n-1)t_n$, we execute Step 3.
- We are sure that N is smaller than $(n-1)t_n$, we go to Step 2.

step 2: we have generated the first p bits of a number N

```
d = n − 1
j = 1
do
    // computation done with 2p bits and rounded to p bits
    N = N − d t_d t_{n−jd}
    if N ≤ 0 then
        break
```

```
        if $j < \lfloor \frac{n-1}{d} \rfloor$ then
                j = j+1
        else
                d = d-1
                j = 1
    while $d > 0$
    if $N > 0$ then
        go to step 3
    else
        return the pair $(j, d)$
```

step 3: this step is the same as Step 2 except that this time we do the computations exactly.

```
    compute the value of $t_i$ for $i$ in $[1, n]$
    finish the drawing of $N$
    if $N \geq (n-1)t_n$ then
            goto step 1
    d = $n - 1$
    j = 1
    do
        // computation done with 2p bits and rounded to p bits
        $N = N - d\,\underline{t_d t_{n-jd}}$
        if $N \leq 0$ then
                break
        if $j < \lfloor \frac{n-1}{d} \rfloor$ then
                j = j+1
        else
                d = d-1
                j = 1
    while $d > 0$
    if $N \leq 0$ then
        return the pair $(j, d)$
    d = $n - 1$
    j = 1
    do
        // computation done with 2p bits and rounded to p bits
        $N = N + d\,\underline{t_d t_{n-jd}}$
        // computation done exactly
        $N = N - d\,t_d t_{n-jd}$
        if $N \leq 0$ then
                break
```

```
        if j <⌊(n-1)/d⌋ then
                j = j+1
        else
                d = d-1
                j = 1
    while true
    return the pair (j, d).
```

9.3.3 The complexity of the algorithm

Since in the preprocessing step all computations are done using numbers with $O(ln(n))$ bits, the real complexity of this step is in $O(n^2\, ln(n)Ln_2(n)Ln_3(n))$.

Now we can estimate the probability that a choice of a pair (j, d) stops because of a problem of accuracy. For each drawing, we need to draw a number N and to compare it with at most $n\,ln(n)$ other integers. Therefore there are at most $n\,ln(n) + 1$ intervals in which N can reside and the boundaries of each interval are know with an accuracy of $\frac{2}{n^4}$. Thus the probability that a drawing stops is at most :

$$(n\,ln(n) + 2)\frac{2}{n^4} \leq 3\frac{ln(n)}{n^3}.$$

Now we use the fact that we must do at most as many choices of a pair (j, d) as there are nodes in the arborescence that we want to draw (ie. n). Therefore, the probability that we need to compute the exact values of the t_i is less than

$$n\,3\frac{ln(n)}{n^3} = 3\frac{ln(n)}{n^2}.$$

Therefore, the average complexity of the recursive part is in

$$O(n\,ln(n)ln(n)Ln_2(n)Ln_3(n) + 3\frac{ln(n)}{n^2}n^3 ln(n)Ln_2(n)) =$$
$$O(n\,ln^2(n)Ln_2(n)Ln_3(n))$$

since in the normal case, we need to do $O(n\,ln(n))$ computations which require a time in $ln(n)Ln_2(n)Ln_3(n)$, and with a probability $3\frac{ln(n)}{n^2}$ the drawing requires a complexity in $O(n^3 ln(n)Ln_2(n))$.

In conclusion, we have obtained an algorithm which is composed of

- a preprocessing step whose complexity is in $O(n^2 ln(n) Ln_2(n) Ln_3(n))$,
- a recursive part whose average complexity is in $O(n\, ln^2(n) Ln_2(n) Ln_3(n))$.

This shows an improvement of a factor $\frac{n}{ln(n) Ln_3(n)}$ over the basic recursive algorithm.

9.4 CONCLUSION

We have seen in this chapter how to draw some trees even when only some coefficients are known and with a low accuracy. Of course, in some cases we need to compute exactly these coefficients. But what it is really interesting is that such a method can decrease greatly the average complexity of some algorithms.

This chapter ends the presentation of the sequential algorithms for generating some trees. We see in the next chapter how to generate some words which are in 1-1 correspondence with forests of trees split into patterns on a parallel computer.

10

A PARALLEL ALGORITHM FOR THE GENERATION OF WORDS

Abstract

A parallel algorithm is presented for generating a permutation of size n. The algorithm uses $O(n)$ processors and runs in $O(Log^2(n))$ time. We show also that this algorithm permits the generation of a Dyck word. The same techniques work for Motzkin words, left factors of Dyck, Motzkin words and words of the language $\mathcal{E}$ (see Chapter 5).

10.1 INTRODUCTION

We present here some methods that can be used to generate, on parallel machines, words that are in 1-1 correspondence with various classes of trees or forests of trees.

The organization of the chapter is as follows : the remainder of this section provides the principles of our parallel algorithm for the generation of a Dyck word. Section 2 describes the generation of a permutation. The generation of a Dyck word is detailed in Section 3 while Section 4 provides extensions to Motzkin words, Dyck left factors, Motzkin left factors, words of language $\mathcal{E}$. Our conclusions and further aspects are offered in Section 5.

Now let us see how the algorithm which generates a Dyck word works, this will explain how the generation of a permutation can be used to generate other structures. This algorithm has four steps.

- generation of a sequence of $2n+1$ integers such that the ith term is between 1 and i,
- transformation of this sequence into a permutation of length $2n+1$,
- transformation of the permutation into a sequence of $n+1$ letters x and n letters y,
- transformation of this sequence into a Dyck word of length $2n$.

These steps have different complexities.

- the first step is simple : We need $2n+1$ processors labelled from 1 to $2n+1$ and let the ith processor choose randomly a number between 1 and i.
- the second step is more intricate : $O(n)$ processors are used and the time complexity is in $O(Log^2(n))$. It is based on a bijection which transforms a sequence of $2n+1$ integers into a permutation [51]. This bijection is then coded on a multiprocessor machine using techniques developed by Batcher [12],[17] and Stone [49]. For simplicity of presentation, we give here an algorithm which uses $O(nLog(n))$ processors, but this algorithm can also be implemented very neatly in time $O(Log^2(n))$ using $O(n)$ processors [54].
- in Step 3, the values of the permutation greater than n are replaced by the letter x and the others by y. The time complexity is in $O(1)$.
- we apply the cycle lemma [21] in parallel. This step needs $O(n)$ processors and a time complexity in $O(Log^2(n))$.

10.2 GENERATION OF A PERMUTATION

Definition 39 *We call lower-exceeding sequence a sequence of integers* $(s_1, s_2, \ldots, s_n)$ *such that*

$$\forall i, 1 \leq s_i \leq i.$$

We recall first how the classical bijection [51] between permutations and lower-exceeding sequences works, then we show how to implement it on a parallel machine with an average time complexity in $O(Log^2(n))$.

10.2.1 Bijection $\mathcal{H}$ between permutations and lower-exceeding sequences

The mapping $\mathcal{H}$ works as follows :

- it transforms a permutation of $\mathcal{S}_n$ into an $n \times n$ matrix [16] such that each row and column contains a circle. We put a circle in the box (i, j) if and only if the permutation transforms i into j.

  Example :

  The permutation $(1, 3, 5, 4, 2)$ gives

  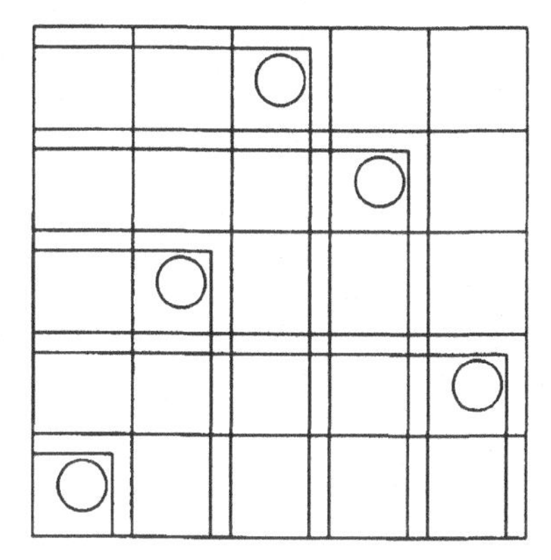

- the mapping then transforms this array into a lower-exceeding sequence. This is easily done by locating the circle in position (i, j) and by associating the number of circles located in the South-West quarter (with respect to the box (i, j)) to the ith term of our sequence.

  Example :

  The preceding array is transformed into

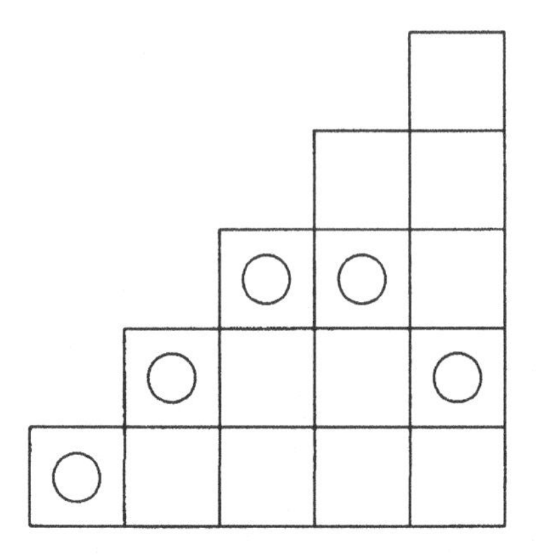

this gives the lower-exceeding sequence $(1, 2, 3, 3, 2)$.

Now we define in a more formal way this mapping $\mathcal{H}$.

Definition 40 *Let $w = (w_1, w_2, \ldots, w_j)$ be a permutation, then $s = (s_1, s_2, \ldots, s_j) = \mathcal{H}(w)$ if and only if*

$$\forall i,\ s_i = card(\{1 \le k \le i \text{ such that } w_k \le w_i\}).$$

In the next subsection we prove that $\mathcal{H}$ is a bijection by constructing its reciprocal mapping.

10.2.2 Definition of $\mathcal{F} = \mathcal{H}^{-1}$

Before defining the function $\mathcal{F}$, we state a theorem :

Theorem 37 *Let $w = (w_1, w_2, \ldots, w_j)$ be a permutation, i an integer of $[\![1, j]\!]$ and $s_l = card(\{1 \le k \le i, w_k \le w_i\})$. Then w_i is the s_l^{th} smallest element of $\mathbb{N}^\star - \{w_{i+1}, w_{i+2}, \ldots, w_j\}$.*

Proof. Since w is a permutation, we can write :

$$\begin{aligned} w_i &= card(\{1 \le k \le j, w_k \le w_i\}) \\ &= card(\{1 \le k \le i, w_k \le w_i\}) + card(\{i < k \le j, w_k \le w_i\}) \\ &= s_l + card(\{i < k \le j, w_k < w_i\}). \end{aligned}$$

Now let q be the s_l^{th} smallest element of $\mathbf{N}^\star - \{w_{i+1}, w_{i+2}, \dots, w_j\}$, we have

$$q = card(\{v \in \mathbf{N}^\star, v \leq q\})$$
$$= card(\{v \in \mathbf{N}^\star - \{w_{i+1}, \dots, w_j\}, v \leq q\}) + card(\{v \in \{w_{i+1}, \dots, w_j\}, v \leq q\})$$
$$= s_l + card(\{i < k \leq j, w_k < w_i\}) = w_i.$$

□

Definition 41 *Consider a set of positive integers* $s = (s_1, s_2, \dots, s_j)$, *then*

$$\mathcal{F}(s) = (w_1, w_2, \dots, w_j)$$
$$\Leftrightarrow \forall i \leq j, w_i \text{ is the } s_i{}^{th} \text{ element of } \mathbf{N}^\star - \{w_{i+1}, w_{i+2}, \dots, w_j\}.$$

Now we have the following theorem:

Theorem 38 *The function* $\mathcal{F}$ *defines a bijection between the permutations of size* n *and the lower-exceeding sequences of size* n. $\mathcal{H}$ *is the inverse mapping of* $\mathcal{F}$.

Proof. From Definition 40 and Theorem 37, we deduce that for all permutations w :

$$\mathcal{F}(\mathcal{H}(w)) = w.$$

The function $\mathcal{F}$ is a bijection because the number of permutations of size n is the same as the number of lower-exceeding sequences of size n (ie. $n!$). □

10.2.3 Implementation of $\mathcal{F}$

The goal of this section is to provide a way to implement correctly the function $\mathcal{F}$ such that it runs in $O(Log^2(n))$ steps. For this purpose we define a function $\mathcal{G}$ which is a little bit more sophisticated than $\mathcal{F}$, this function will be easy to code on a parallel machine and will give a result as useful as the result of $\mathcal{F}$.

Definition of $\mathcal{G}$

Definition 42 *Let* $s = (s_i, s_{i+1}, \dots, s_j)$ *be a sequence of integers and* $w = (w_i, \dots, w_j) = \mathcal{F}(s)$, *then*

$\mathcal{G}(i,s) = ((x_i, a_i), \ldots, (x_j, a_j))$ *if and only if*

- $x_i < x_{i+1} < \cdots < x_j$,
- $\forall\, i \leq k \leq j, i \leq a_k \leq j$ *and* $x_k = w_{a_k}$.

Remark :

In other words, $\mathcal{G}(i, s)$ returns the sorted list of (second) coordinates of boxes of the array $n \times n$ which contain the circles.

Investigation of $\mathcal{G}$

Assume that we know :

- $\mathcal{G}(i, (s_i, \ldots, s_j)) = ((x_i, a_i), \ldots, (x_j, a_j))$,
- $\mathcal{G}(j+1, (s_{j+1}, \ldots, s_l)) = ((x_{j+1}, a_{j+1}), \ldots, (x_l, a_l))$,

and that we want to compute $\mathcal{G}(i, (s_i, \ldots, s_l)) = ((y_i, b_i), \ldots, (y_l, b_l))$.

Lemma 5 *We get then*

$$\forall\, k \in \,]\, j, l\,]\,, \exists\, m \in [\, i, l\,]\,, (x_k, a_k) = (y_m, b_m).$$

Proof. In fact let

- $\mathcal{F}((s_{j+1}, \ldots, s_l)) = (w_{j+1}, \ldots, w_l)$,
- $\mathcal{F}((s_i, \ldots, s_l)) = (v_i, \ldots, v_l)$.

By definition $x_k = w_{a_k}$. Since w_n (respectively v_n) is defined by the values of $s_n, \ldots, s_l$, we have $w_k = v_k$ if k is greater than j. This implies that : $x_k = v_{a_k}$.

Coming back to the definition of $\mathcal{G}$, we see that there exists a pair (y_m, b_m) with $y_m = x_k$ and $b_m = a_k$. □

Now we show what happens if k is between i and j.

Lemma 6 *For each k such that $i \leq k \leq j$, denote by z_k the $x_k{}^{th}$ element of $N^\star - \{x_{j+1}, x_{j+2}, \ldots, x_l\}$. Then there exists an integer v such that $(y_v, b_v) = (z_k, a_k)$.*

Proof. Let

- $\mathcal{F}((s_i, \ldots, s_j)) = (w_i, \ldots, w_j)$,
- $\mathcal{F}((s_i, \ldots, s_l)) = (v_i, \ldots, v_l)$.

By definition we have $x_k = w_{a_k}$. We use now Definition 41 in order to compute v_{a_k},

$$\begin{aligned} v_{a_k} &= s_{a_k} + card(\{m \text{ such that } a_k < m \leq l \text{ and } v_m \leq v_{a_k}\}) \\ &= s_{a_k} + card(\{m \text{ such that } a_k < m \leq j \text{ and } v_m \leq v_{a_k}\}) \\ &\quad + card(\{m \text{ such that } j < m \leq l \text{ and } v_m \leq v_{a_k}\}) \\ &= N + card(\{m \text{ such that } j < m \leq l \text{ and } v_m \leq v_{a_k}\}) \end{aligned}$$

where $N = s_{a_k} + card(\{m \text{ such that } a_k < m \leq j \text{ and } v_m \leq v_{a_k}\})$.

Since the numbers v_m are pairwise distinct, v_{a_k} is the Nth element of $\mathbf{N}^\star - \{v_{j+1}, \ldots, v_l\} = \mathbf{N}^\star - \{x_{j+1}, x_{j+2}, \ldots, x_l\}$.

It remains to prove that $N = x_k$. We choose now an integer m such that $a_k < m \leq j$ and f the only increasing bijection from $\mathbf{N}^\star - \{w_{m+1}, \cdots, w_j\}$ to $\mathbf{N}^\star - \{v_{m+1}, \cdots, v_l\}$. The integer w_m is then the $s_m{}^{th}$ element of the first set and v_m the $s_m{}^{th}$ element of the second, therefore $f(w_m) = v_m$. By induction we get $f(w_{m'}) = v_{m'}$ for all integers $m' \leq m$. This is true in particular if $m' = a_k$, therefore $f(w_{a_k}) = v_{a_k}$. Using the fact that f is increasing, we get :

$$w_{a_k} > w_m \Leftrightarrow f(w_{a_k}) > f(v_m) \Leftrightarrow v_{a_k} > v_m.$$

and

$$N = s_{a_k} + card(\{m \text{ such that } a_k < m \leq j \text{ and } w_m \leq w_{a_k}\}) = x_k.$$

□

Theorem 39 *Let m and n be two integers between i and l and consider the integers o and p such that $a_m = b_o$ and $a_n = b_p$. Then $o < p$ if and only if :*

- $m \leq j$ *and* $n \leq j$, $m < n$,
- $m > j$ *and* $n > j$, $m < n$,
- $m \leq j$ *and* $n > j$, $x_m + (n - j) \leq x_n$,
- $m > j$ *and* $n \leq j$, $x_n + (m - j) \leq x_m$.

Proof. The two first relations follow directly from Lemmas 5 and 6. The two last cases are equivalent if we replace n by m, we can therefore restrict our study to the case where $m \leq j$ and $n > j$.

Assume that $m \leq j$ and $n > j$, then :

- y_p is equal to x_n,
- y_o corresponds to the ${x_m}^{th}$ element of $\mathbf{N}^\star - \{x_{j+1}, \ldots, x_l\}$.

The inequality $y_p > y_o$ is satisfied if and only if there are more than x_m elements in $[\![1, x_n [\![- \{x_{j+1}, x_{j+2}, \ldots, x_{n-1}\}$ (i.e if and only if $x_m \leq x_n - 1 - (n - 1 - j)$). □

This theorem provides a way to compute the new position of the image of a pair (x_p, a_p) in $\mathcal{G}(1, (s_1, \ldots, s_l))$ by doing only comparisons between the pairs (x_m, a_m) and (x_n, a_n). Assume, in addition, that we succeed in reordering the pairs (x_m, a_m) with the order induced by Theorem 39. We can then go quickly from this representation to $((y_i, b_i), \ldots, (y_l, b_l))$. In fact, it is sufficient to transform all pairs (x_m, a_m) according to the following rule :

Theorem 40 *Let n be the new position of the pair (x_m, a_m),*

- *if* $a_m > j$ *then* $(y_n, b_n) = (x_m, a_m)$,
- *if* $a_m \leq j$ *then* $(y_n, b_n) = (x_m + n - m, a_m)$.

Proof. If $a_m > j$, we use Lemma 5 in order to conclude.

Assume now that $a_m \leq j$, we have therefore :

$n - i + 1 = card(\{i \leq k \leq n\})$
$= card(\{i \leq k \leq n, y_k \leq y_n\})$
$= card(\{i \leq k \leq n, b_k \leq j \text{ and } y_k \leq y_n\}) +$
$card(\{i \leq k \leq n, b_k > j \text{ and } y_k \leq y_n\})$
$= card(\{i \leq k \leq j, x_k \leq x_m\}) +$
$card\left(\{k > j \text{ such that } x_k \text{ is less than the } x_m^{th} \text{ element of } \mathbf{N}^\star - \{x_{j+1}, \ldots, x_l\}\}\right)$
$= m - i + 1 +$
$card\left(\{k > j \text{ such that } x_k \text{ is less than the } x_m^{th} \text{ element of } \mathbf{N}^\star - \{x_{j+1}, \ldots, x_l\}\}\right).$

but using Lemma 6, we get :

$y_n = \text{ the } x_m^{th} \text{ element of } \mathbf{N} - \{x_j, \ldots, x_l\} = x_m +$
$card\left(\{k > j \text{ such that } x_k \text{ is. less than the } x_m^{th} \text{ element of } \mathbf{N}^\star - \{x_{j+1}, \ldots, x_l\}\}\right)$

Therefore $y_n = x_m + n - m$. □

Implementation of the algorithm

The generation of a permutation is done as follows : we generate first a lower-exceeding sequence $(s_1, \ldots, s_n)$ with the help of n processors. The ith processor chooses a random number in the interval $[1, i]$. Then, using a network which we will describe later, we compute $\mathcal{G}(1, (s_1, \ldots, s_n)) = ((1, a_1), \ldots, (n, a_n))$. We get a permutation $(a_1, \ldots, a_n)$.

Theorem 41 *All permutations $(a_1, \ldots, a_n)$ of size n are obtained with probability $\frac{1}{n!}$.*

Proof. The permutation obtained with our algorithm is in fact the inverse of the permutation $\mathcal{F}((s_1, \ldots, s_n))$. The probability of generating a permutation w is therefore equal to the probability of generating, at the beginning, a lower-exceeding sequence $\mathcal{F}^{-1}(w^{-1}) = \mathcal{H}(w^{-1})$, this means $\frac{1}{n!}$. □

It remains to find a processor network which is able to compute the function $\mathcal{G}$. In fact, we can take all processor networks which sort a list of n elements by implementing the merging sort. Indeed, if we have a processor which can merge two sorted lists of size k and if we know $\mathcal{G}(i,(s_i,\ldots,s_{i+k-1}))$, $\mathcal{G}(i+k,(s_{i+k},\ldots,s_{i+2k-1}))$, then this processor computes the value of $\mathcal{G}(i,(s_i,\ldots,s_{i+2k-1}))$.

We assume for simplicity that $n = 2^p$, and that we have a processor $\mathcal{N}(p)$ which can merge two sorted lists of 2^{p-1} elements. The following network does the job :

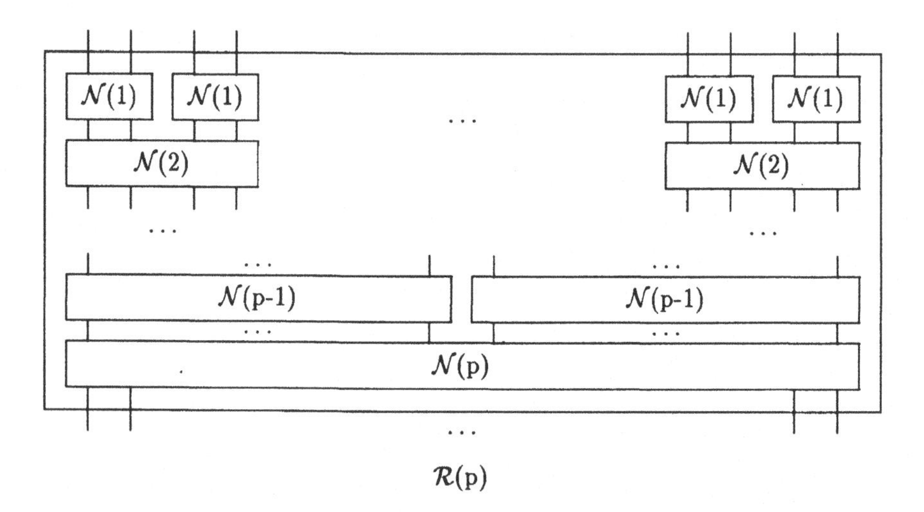

$\mathcal{R}$(p)

where $\mathcal{N}(P)$ is a network which is able to merge two sorted lists with the help of Theorem 39. Then it uses the results of Theorem 40 in order to get the value of $\mathcal{G}$.

There exists a network which is able to merge two sorted lists of size 2^{p-1} in time $O(p)$ with approximately 2^p processors (see [12], [49] for more details. If we use this network, we have a full network with $O(n\ Log(n))$ processors but this network can be neatly implemented with $O(n)$ processors using a shuffle-exchange network [54]. Such a network transforms a lower-exceeding sequence s into $\mathcal{F}(s)^{-1}$, and solves our problem in time $O\left(Log^2(n)\right)$.

10.3 GENERATION OF A DYCK WORD

We have shown in the previous section how to get a random permutation of size $2n+1$. We replace now in this permutation the numbers strictly greater than n by the letter x and the others by the letter y in order to get a random word with $n+1$ letters x and n letters y distributed among $2n+1$ processors.

We show here how to transform it into a 1-dominated sequence (i.e. a letter x followed by a Dyck word). Let $u_0, u_1, \ldots, u_{2n}$ be the word composed of letters x and y and define the functions :

- $f(i,j) = card(\{i \leq l \leq j, u_l = x\}) - card(\{i \leq l \leq j, u_l = y\})$,
- $g(i,j)$ the last position where the minimum of the function $v \mapsto f(i,v)$ appears in the interval $[\, i,j]$,
- $h(i,j)$ the value of the minimum of the function $x \mapsto f(i,x)$ in the interval $[\, i,j]$.

The value $g(0,2n)$ permits us to know where to perform the cyclic transformation which will transform the word composed of x's and y's into a 1-dominated word.

10.3.1 Computation of $g(0,2n)$

The function g is strongly related to f and h and their computation will be done simultaneously. The computation is by induction. In fact :

Proposition 23 *For f, g and h when $i < j < l$:*

- $f(i,i) = 1$ *if* $u_i = x$ *and* $f(i,i) = -1$ *if* $u_i = y$,
- $g(i,i) = i$,
- $h(i,i) = f(i,i)$,
- $f(i,l) = f(i,j) + f(j+1,l)$,
- $h(i,l) = min\,(h(i,j), f(i,j) + h(j+1,l))$,
- $g(i,l) = \begin{cases} g(i,j) & \text{if } h(i,j) < f(i,j) + h(j+1,l), \\ g(j+1,l) & \text{if } h(i,j) \geq f(i,j) + h(j+1,l). \end{cases}$

Proof. The three first properties follow directly from the definitions of f, g, h. The proof of the other properties is by induction. □

This property permits recursive computation of the values of f, g, h. We will describe a network using $O(n)$ processors which does the job in time $O(Log(n))$. For simplicity we assume that $2n+1$ is a power of 2 (this assumption will not change the following results). The network is a perfect binary tree with $2n+1$ leaves which are the processors containing the terms of the sequence $(u_0, u_1, \ldots, u_{2n})$. The values of f, g and h are computed for each leaf, then the values of f, g and h are computed with the parameters $(2k, 2k+1)$, $(4k, 4k+3)$, $\ldots$, $(0, 2n)$ one step bottom up in the tree.

Example :

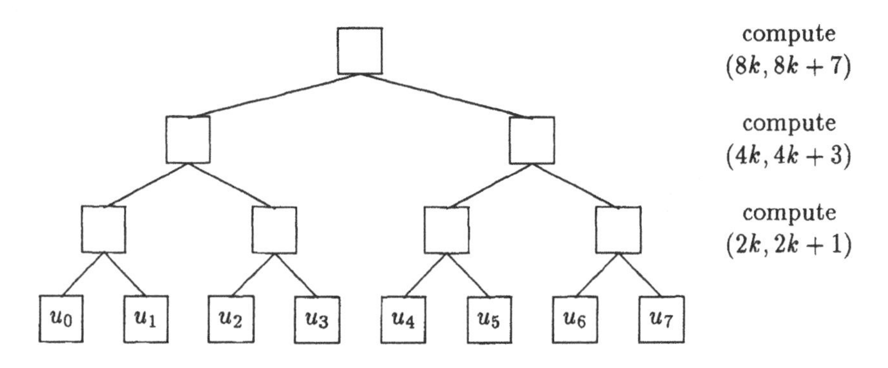

As soon as the value of $g(0, 2n)$ is computed, we can propagate this result in time $O(Log(n))$ so that each processor knows where the cyclic permutation has to be performed.

10.3.2 Cyclic permutation

We apply the cycle lemma 2 (page 76), with the value $N = g(0, 2n)$ as pivot. This brings the terms $(u_{N+1}, \ldots, u_{2n})$ into position $(u_0, \ldots, u_{2n-N-1})$ and the terms $(u_0, \ldots, u_N)$ into position $(u_{2n-N}, \ldots, u_{2n})$ and can be done with a sorting network, because we know in which places the values of the elements have to be. These numbers are used as keys on which we will do the sorting. This leads to an algorithm whose time complexity is in $O(Log(n))$ if we use the sorting network proposed in [3] or in $O(Log^2(n)$ with a merge sorting network [49]. These are $O(n)$ processors networks.

10.4 EXTENSIONS

The techniques developed apply also to the parallel generation of a Motzkin word, a Dyck left factor, a Motzkin left factor or a word of the language $\mathcal{E}$. We show below how to design a parallel algorithm for generating these classes of words.

10.4.1 A parallel algorithm for the generation of a Motzkin word

We parallelize the Marty sequential algorithm that we have presented in Subsection 7.2.3.

First we choose a sequence of $\lceil \frac{2n}{3} \rceil$ adjacent bits. This needs $t = \lceil \frac{2n}{3} \rceil$ processors, each of them chooses randomly a bit ;

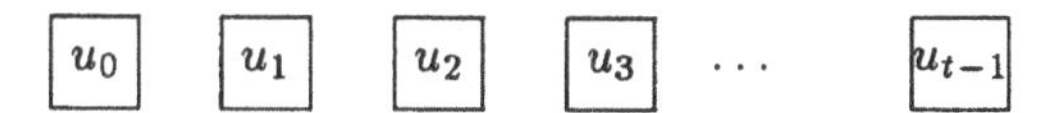

Then we count the number k of bits 1 which have been chosen. This is done in time $O(Log(n))$ on $O(n)$ processors with the help of the perfect binary tree network with $\lceil \frac{2n}{3} \rceil$ leaves :

Example :

If $t = 8$, we use the following network :

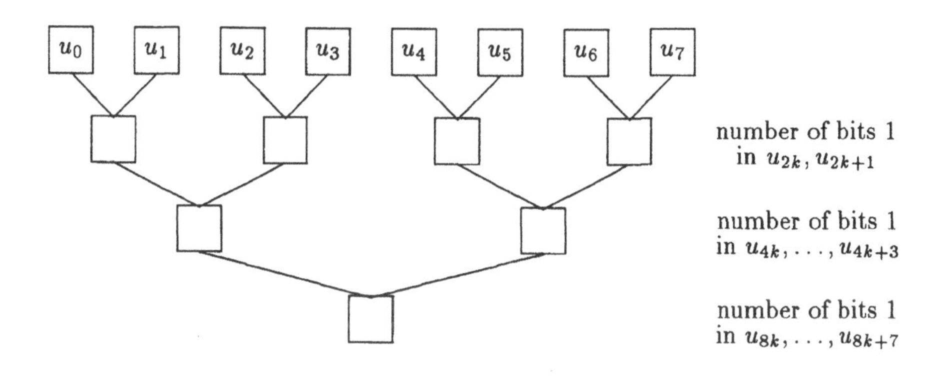

Then we must verify if the value is accepted or rejected. This is clear if $k = 0$ or when $k > \lfloor \frac{n+1}{2} \rfloor$.

In the other case, we compute the corresponding integers a, b and c defined in the Proposition 14 (page 123) in time $O(1)$ with a single processor. Then c processors are informed that they have to participate in this verification (a classical binary tree structure is used here). This needs $O(c)$ processors and a time in $O(Log(c))$. The ith processor chooses randomly a number in $[\![1, b+1-i]\!]$ and considers that its drawing succeeds if $n \leq a+1-i$. The global result is then recovered thanks to the tree structure which does now the "*and*" of the values given by the c processors in time $O(Log(c)) = O(Log(n))$ with $O(c) = O(n)$ processors.

In case of failure, we start over with the same step (we can inform the $\lceil \frac{2n}{3} \rceil$ processors at the beginning in time $O(Log(n))$ thanks to a binary tree structure with $O(n)$ processors). In case of success, we execute the next step. An attempt for choosing the integer k requires therefore a time in $O(Log(n))$ and $O(n)$ processors. The total average time for executing this step is therefore also in $O(Log(n))$.

The second step generates a Motzkin word of size $n-1$ with $k-1$ letters x. This is done in time $O(Log^2(n))$ on $O(n)$ processors using similar methods as for the generation of a Dyck word.

First, we generate a random permutation of size n. Then we replace the numbers of this permutation that are smaller or equal to k by the letter x, thoses which are greater than $2k$ by the letters a and the remaining ones by the letters y. Finally, we apply on this word the cyclic permutation that transforms it into a letter x followed by a Motzkin word.

10.4.2 Generation of a left factor of Dyck or a Motzkin word

Generation of a Dyck left factor

We use the 1-1 correspondence that exists between the Dyck left factors with n letters and the sequences of $\lceil \frac{n}{2} \rceil$ letters x and $\lfloor \frac{n}{2} \rfloor$ letters y (see Theorem 29 page 128).

Therefore, we generate a permutation of size n. Then we replace the values of this permutation which are strictly greater than $\lceil \frac{n}{2} \rceil$ by y and the others by x.

Finally, we compute the height $f(0,i)$ of each letter as for the generation of a Dyck word; and we transform a letter $u_i = x$ (resp. $u_i = y$) into y (resp. x) if and only if $f(0,i) < 0$.

Generation of a Motzkin left factor

We proceed in the same way as for the generation of a Motzkin word but this time we transform the Denise algorithm (see the Subsection 7.3.5). The first steps of this algorithm consist in choosing the number of letters x and y that our word will have; these steps are parallelized in the same way as for the parallelization of the Marty algorithm described above.

Finally, we must draw a Motzkin left factor with k letters a. We proceed as follows. First, we generate a permutation of n elements. Then we replace the values of this permutation which are in $[\![1, \lceil \frac{k}{2} \rceil]\!]$ by x, those which are in $[\![\lceil \frac{k}{2} \rceil + 1]\!]$ by y and the remaining values by a. Finally, we compute the values of $f(0,i)$ and transform a letter $u_i = x$ (resp. $u_i = y$) into y (resp. x) if and only if $f(0,i) < 0$.

10.4.3 Generation of a word of $\mathcal{E}$

We present in this paragraph a method to draw a word of $\mathcal{E}$, these words are in 1-1 correspondence with the forest of trees split into patterns (see the Chapter 5).

Indeed, if we denote by $\mathcal{M} = ((M_1, a_1), \cdots, (M_k, a_k))$ a multiset of patterns and if we want to generate a word of the language $\mathcal{E}$ that is in 1-1 correspondence with the forest of p trees split into patterns $(F, \mathcal{M}, f)$, we can parallelize the main steps of the sequential algorithm. This gives the following algorithm :

- First, we generate a random permutation of size $s = \sum_{i=1}^{k} a_i$. Then we replace the numbers that are between 1 and a_1, between $a_1 + 1$ and $a_1 + a_2$, ..., between $s - a_k + 1$ and s by the code of the patterns M_1, ..., M_k. Thus we obtain a word v.
- We generate a random permutation of size $n - p - e - c + d - 1$, we replace the numbers smaller than d by 0 and the others by 1. Then we insert

between the ith symbols () j letters f where j is the number of 1's that are between the $(i-1)^{th}$ and the ith letter 0 of the obtained sequence. We obtain a new word w.

- Finally, we compute the minimal height h of the symbols of w, we draw a random number in $[1,p]$. Then we find the last letters x, x_j, x_j^j of w which end with height $h+r$ and we perform the cyclic permutation that puts this letter at the beginning of the resulting word.

When the patterns M_1, ..., M_k are chosen, all the steps of this algorithm can be implemented to obtain an algorithm whose complexity is in $O(Log^2(n))$ on a computer with $O(n)$ processors by using the same methods than for the generation of a Dyck word.

10.5 CONCLUSION

We have shown how to generate a permutation of size n, a Dyck word, a Dyck left factor, a Motzkin word, a Motzkin left factor and a word of the language $\mathcal{E}$ with $O(n)$ processors in time $O(Log^2(n))$. This time is indeed only the average time when we generate a Motzkin word or a Motzkin left factor.

Unfortunately it would be very difficult to use this approach to parallelize the sequential algorithms that are based on recursion (for instance the RANRUT algorithm).

APPENDIX 1

HORTON-STRAHLER'S NUMBERS

The content of this section is from [28],[52].

The hydrogeologist Horton [32] proposed a method for classifying the rivers of a fluvial network. This technique was later improved by Strahler [50]. We present in this section the principles of their approach: the order of a river issued from the source is 0, two confluent rivers of order k give rise to a river of order $k+1$, a river of order i confluent with a river of order k $(i < k)$ gives rise to a river of order k.

Roughly speaking, the order of a river reflects its importance inside the fluvial network. A segment of order k is, by definition, the maximal portion of a river of order k. In other words, a segment of order k starts at the junction of two segments of order $k-1$ (if $k=0$, it starts at the source) and finishes at the intersection of a river of order $k' > k$ (if k is the maximal order, the segment ends in the sea!).

Let b_k be the number of segments of order k. The branching ratio β_k of order k is defined by: $\beta_k = \frac{b_k}{b_{k+1}}$.
Hydrogeologists observed that for a given fluvial network, β_k is a constant β whose value is between 3 and 5.
For a given fluvial network, we can define the average length l_k of the segments of order k. It has been observed that, in general, the ratio $\frac{l_{k+1}}{l_k}$ is a constant approximately equal to $\frac{\beta}{2}$.

The Strahler number of a binary tree is defined recursively as follows:

- 0 for a leaf,

- the number $l+1$ for a node whose two children have Strahler numbers equal to l,
- $max(l_1, l_2)$ for a node whose two children have Strahler numbers equal to l_1 and l_2 ($l_1 \neq l_2$).

The Strahler number of a binary tree is the maximal order of its nodes.

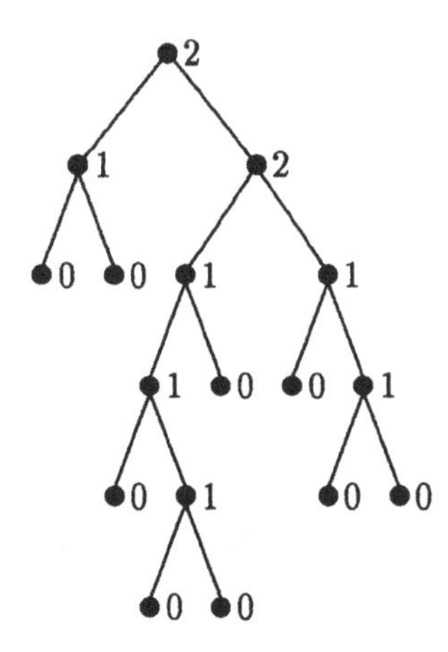

Let B be a binary tree whose Strahler number is equal to k and let x be an internal node. The biorder of x is the pair of orders of the edges joining x to its children.

We define successively the following quantities:
a_k the number of internal nodes whose order is equal to k,
$b_{k,i}$ the number of nodes whose biorder is equal to (k, i) with $0 \leq i < k$,
$b_{k,k}$ the number of nodes whose biorder is equal to $(k-1, k-1)$with $1 \leq k$,
$p_{k,j} = \frac{b_{k,j}}{a_k}$ for $1 \leq k$,
The number $p_{k,j}$ is the probability for a node of order k to have a biorder equal to (k, j).
The ramification matrix $R(B)$ of the binary tree B is the stochastic matrix $R(B) = (p_{k',j})$ where $0 \leq j \leq k' \leq k$ and $1 \leq k' \leq k$.
$R(B)$ gives information about the shape of the tree B.

Example :

Observe the two extreme cases:

- the vine-branch case

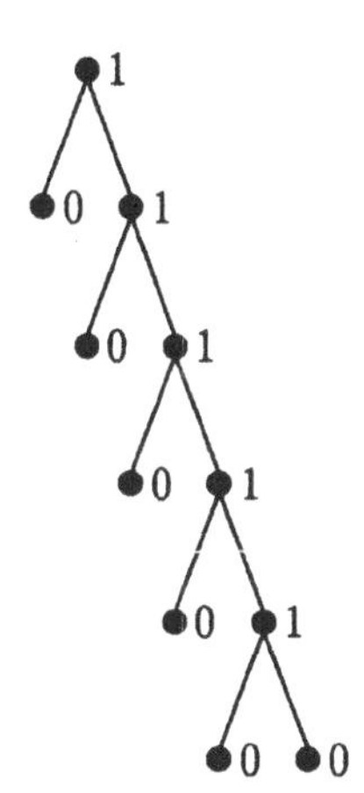

whose ramification matrix is

$p_{k,i}$	0	1
1	$\frac{4}{5}$	$\frac{1}{5}$

- the perfect tree case

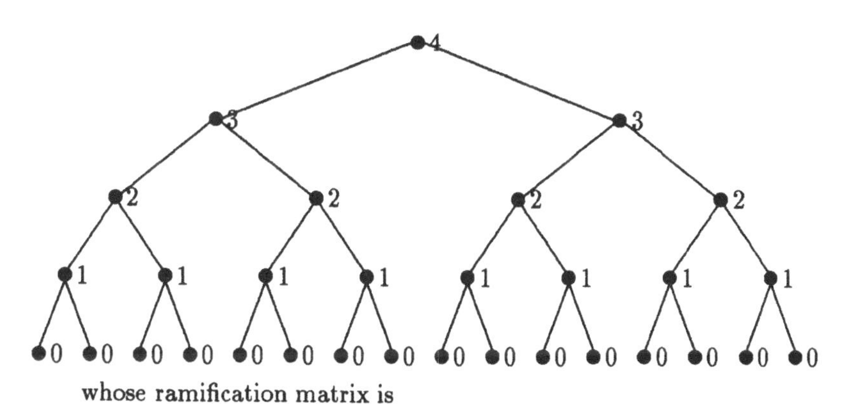

whose ramification matrix is

$p_{k,i}$	0	1	2	3	4
1	0	1			
2	0	0	1		
3	0	0	0	1	
4	0	0	0	0	1

D. Arquès, G. Eyrolles, N. Janey, G. Viennot (cf. [24] and [1]) used Strahler numbers and ramification matrices in order to generate tree structures. They use the length, the thickness of each edge, deviation and branching angles (see Figures below).

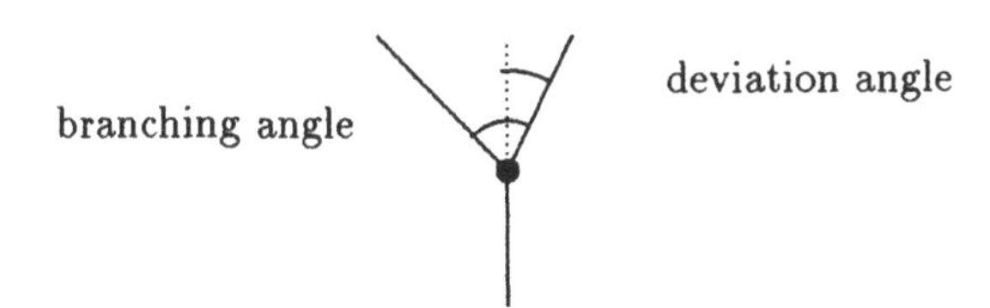

The length and the thickness of each edge depend only on its order in the tree. Deviation and branching angles depend only on the corresponding biorder.

With the help of these techniques applied to different varieties of trees, we produced the pictures of this volume.

APPENDIX 2

ALGORITHMS

We present below the C code of two algorithms for the random generation of:
i) binary trees,
ii) unary-binary trees
of size n.

2.1 GENERATION OF BINARY TREES : RÉMY'S ALGORITHM

```
#define N 1000

typedef struct {
    int right_child, left_child, parent;
    int num;
} Node;

Node tree [2*N+1];

// construction of a growing binary tree with n internal nodes (n ≤ N)

// modifies the position of two leaves, does not modify the tree built

void change_leaves (int a, int b)
{
    int parenta, parentb;
    parenta = tree[a]->parent;
```

```
    parentb = tree[b]->parent;
    if (tree [parenta]->right_child == a)
      tree [parenta]->right_child = b;
    else
      tree [parenta]->left_child = b;
    tree[a]->parent = parentb;
    if (tree [parentb]->right_child == b)
      tree [parentb]->right_child = a;
    else
      tree [parentb]->left_child = a;
    tree[b]->parent = parenta;
}

void growing_tree (long n)
{
    long i, number, tmp;

    // built a tree of size 1
    tree [0]->left_child = 1;
    tree [0]->right_child = 2;
    tree [0]->num = 1;
    tree [1]->parent = tree [2]->parent = 0;
    tree [1]->right_child = tree [1]->left_child = -1;
    tree [2]->right_child = tree [2]->left_child = -1;

// the internal nodes will be in the boxes [ 0, i - 1[  of the tree's array
// the leaves in boxes [ i, 2 * i[  of the tree's array

    for (i = 2; i <= n; i++){
        number = random (i); // random number of [ 0, i[
        // the leaf is in the box number+i-1
        change_leaves (i-1, number+i-1);
        tree [i-1]->right_child = 2*i-1;
        tree [i-1]->left_child = 2*i;
        tree [i-1]->num = i;
        tree [2*i-1]->parent = tree [2*i]->parent = i-1;
        tree [2*i-1]->right_child = tree [2*i-1]->left_child = -1;
        tree [2*i]->right_child = tree [2*i]->left_child = -1;
    }
}
```

2.2 GENERATION OF UNARY-BINARY TREES : SAMAJ LAREIDA'S ALGORITHM

```
#define N 300

typedef struct {
    int number_of_children;
    int children [2];
} Node;
Node tree [N];

// generation of the Motzkin word of size n

// this function generates k integers 0 or 1 and gives the number of 1's generated

long choose_k (long k)
{
    long result = 0;
    while(k--)
        if (random (2) == 1)
          result++;
    return result;
}

// this function returns true with probability (a!/(a-c)!c!) /(b!/(b-c)!/c!)

char accept (long a, long b, long c)
{
    long i;

    for (i = 0; i < c; i++){
        if (random (b--) >= a--)
          return 0;
    }
    return 1;
}

// this function validates the choice of the number of ascending steps
```

```
char accept_ascending_steps (long p, long N)
{
    long n = N/3, k;
    if ((p < 0) || (2*p-1 > N))
      return 0;
    if (p-1 == n)
      return 1;
    if (p-1 > n)
      return accept (p-1-n, p-1, N-n-p);
    else
      return accept (n-p+1, N+1-2*p, n);
}

// this function mixes a symbols 1, b symbols -1 and c symbols 0

void mix (long a , long b, long c, int *result)
{
    long i, n, tmp;
    n = a+b+c;
    for (i = 0 ; i < n ; i++){
    // chooses a random number in [ 1, a + b + c]
            tmp = 1+random (a+b+c);
            if (tmp ≤ a) {
               result [i] = 1;
               a--;
            }
            else if (tmp ≤ a+b) {
               result [i] = -1;
               b--;
            }
            else{
               result [i] = 0;
               c--;
            }
    }
}

// finds the conjugate of the searched word which is a Motzkin word
// preceded by an ascending steps

void cyclic_permutation (int *result, int n)
{
```

```
    long height = 0, heightmin = 0, place = 0, i;
    int *stock;

    // finds the first letter of the conjugate word

    for (i= 0; i < n; i++){
        if (result [i] == 1){
          if (height <= heightmin){
            heightmin = height;
            place = i;
          }
        }
        height += result [i];
    }

    // realizes the cyclic permutation

    stock = (int *) malloc (sizeof (int) * n);
    for (i = 0; i < n; i++)
        stock [i] = result [i];
    for (i = 0; i < n; i++){
        result [i] = stock [place++];
        place %= n;
    }

    free (stock);
}


// generates a Motzkin word of size n preceded by an ascending step

void motzkin (long n, int *result)
{
    long number_of_ascending_steps;
    // choose the number of ascending steps
    do
       number_of_ascending_steps = choose_k (n - n/3);
    while (! accept_ascending_steps (number_of_ascending_steps, n));
    // mixes the different letters
    mix (number_of_ascending_steps, number_of_ascending_steps-1,
         n-2*number_of_ascending_steps+1, result);
    // apply the cycle lemma
```

```
    cyclic_permutation (result, n)
}

// transformation of a word comprising an ascending step and a Motzkin word
// into a unary-binary tree

// functions for sharing the stacks

int stackptr = 0; // a pointer on the stack
int stack [N];

void push_node (int node)
{
    stack [stackptr++] = node;
}

int pop_node ()
{
    return stack [--stackptr];
}

// decodes and transforms a Motzkin word of size n into a tree
void decode (int *word, long n)
{
    int i, newnode = 0;
    for (i = 0; i < n; i++){
        tree [newnode].children [0] = tree [newnode].children [1] = -1;
        switch (word [i]){
                case -1 :
                    tree [newnode].number_of_children = 0;
                    break;
                case 0 :
                    tree [newnode].number_of_children = 1;
                    tree [newnode].children [0] = pop_node ();
                    break;
                case 1 :
                    tree [newnode].number_of_children = 1;
                    tree [newnode].children [0] = pop_node ();
                    tree [newnode].children [1] = pop_node ();
                    break;
        }
        push_node (newnode++);
```

```
    }
}

// generate a unary-binary tree of size N

void unary_binary_tree (int N)
{
    int *motzkin_word = malloc (sizeof (int) * N+1);
    motzkin (N+1, motzkin_word);
    decode (motzkin_word, N+1);
    // the result is in the array tree
    // the root is stored in tree [N]
}
```

APPENDIX 3

PICTURES OF TREES

Figure 33.1 A binary tree with 500 nodes

Figure 33.2 A binary tree with 500 nodes

Figure 33.3 A ternary tree with 1000 nodes

Figure 33.4 An increasing tree with 2000 nodes

Figure 33.5 An increasing ternary tree with 600 nodes

Figure 33.6 An increasing ternary tree with 600 nodes

Figure 33.7 An increasing tree with 1000 nodes

REFERENCES

[1] D.Arquès, G. Eyrolles, N.Janey, X.G.Viennot, Combinatorial analysis of ramified patterns and computer imagery of trees, Proc. Siggraph'89, Computer Graphics, 23, 3, 1989.

[2] M.Adel'son-Vel'skiĭ, E.M.Landis, An algorithm for the organization of information, Soviet Mathematics Doklady, 3, 1259-1263, 1962.

[3] M.Ajtai, J.Komlós, E.Szemerédi, An $O(n\,Log(n))$ sorting network, Proceedings of the Fifteenth Annual ACM Symposium on Theory of Computing, 1-9, 1983.

[4] S.G.Akl, I.Stojmenovic, Generating Binary Trees in Parallel, rapport TR-91-22, Université d'Ottawa, May 1991.

[5] L.Alonso, Structures arborescentes : algorithmes de génération, problème de l'inclusion, relations maximin, Thèse, Université de Paris-Sud, Centre d'Orsay, November 1992.

[6] L.Alonso, Uniform generation of a Motzkin word, Theoretical Computer Science (in press)

[7] L.Alonso, R.Schott, Random Generation of Trees with a Given Height, Proceedings of ICCAM'94 (6th International Congress on Computational and Applied Mathematics, Leuven, Belgium, July 1994).

[8] L.Alonso, J.L.Rémy, R.Schott, A linear time algorithm for the generation of trees, Rapport CRIN 90R001, 1990, submitted.

[9] L.Alonso, R.Schott, Random Unlabelled Trees Revisited, Proceedings of ICCI'94 (6th International Conference on Computing and Information, Peterborough, Ontario, Canada, May 1994).

[10] L.Alonso, R.Schott, A parallel algorithm for the generation of a permutation and some applications, Theoretical Computer Science (to appear).

[11] R.Bayer, Symmetric binary B-trees, Data structure and maintenance algorithms, Acta informatica, 1, 290-306, 1972.

[12] K.E.Batcher, Sorting networks and their applications, 1968 Spring Joint Computer Conf., AFIPS Proc., vol. 32. Washington, D.C.: Thompson, pp. 307-314, 1968.

[13] E.Barcucci, R.Pinzani, R.Sprugnoli, The Random Generation of Directed Animals, TCS, vol. 127.2, pp. 333-350, 1994.

[14] B.Chan, S.G.Akl, Generating combinations in parallel, BIT, 26, 2-6, 1986.

[15] G.H.Chen, M.S..Chern, Parallel generation of permutations and combinaisons, BIT, 26, 277-283, 1986.

[16] L.Comtet, Analyse Combinatoire, Presses Universitaires de France, 1970.

[17] T.Cormen, C.Leiserson, R.Rivest, introduction to ALGORITHMS, The MIT Press, Cambridge, England, 1990.

[18] A.Denise, Méthodes de génération d'objects combinatoires de grande taille et problème d'énumération, Thèse, Université de Bordeaux I, Janvier 1994.

[19] N.G.De Bruijn, D.E.Knuth, S.O.Rice, The Average Height of Planted Plane Trees, Graph Theory and Computing, Academic Press, New York, 15-22, 1972.

[20] N.Dershowitz, S.Zaks, Patterns in Trees, Discrete Applied Math. 25, 241-255, 1989.

[21] N.Dershowitz, S.Zaks, The Cycle Lemma and some applications, Europ. J. of Comb. 11, 35-40, 1990.

[22] A. Dvoretzky, Th. Motzkin, A problem of arrangements, Duke Math. J., 24, 305-313, 1947.

[23] M.C.Er, Enumeration Ordered Trees Lexicographically, The Computer Journal, 28, 5, 538-542, 1985.

[24] G.Eyrolles, Synthèse d'images figuratives d'arbres par des méthodes combinatoires, Thèse de troisième cycle, Université de Bordeaux 1, 1986.

[25] P.Flajolet, Mathematical Methods in the Analysis of Algorithms and Data Structures, Rapport Inria n° 400, mai 1985.

[26] P.Flajolet, A.M.Odlysko, The Average Height of Binary Trees and Other-Simple Trees, J. Computer and System Science 25, 171-213, 1982.

[27] P.Flajolet, P.Zimmermann, B.V.Cutsem, A calculus for the Random Generation of Combinatorial Structures, TCS, 29 pages, to appear. Also available as Inria Research Report 1830 (anonymous ftp on ftp.inria.fr dir INRIA/publication/RR file RR-1830.ps.gz).

[28] J.Françon, Arbres et nombres de Strahler dans diverses sciences, Revue du Palais de la Découverte, Paris, 12, 120, 29-36, 1984.

[29] D.Gouyou-Beauchamps, G.Viennot, Equivalence of the Two-Dimensional Directed Animal Problem to a One-Dimensional Directed Animal Problem to a One Dimensional Path Problem, Advances in Applied Math. 9, 334-357, 1988.

[30] R.L.Graham, D.E.Knuth, O.Patashnik, Concrete Mathematics, Addison Wesley, 1990.

[31] P.Gupta, D.T.Lee, C.K.Wong, Ranking and unranking of B-trees, Journal of Alg., 4, 51-60, 1983.

[32] R.E. Horton, Erosioned development of systems and their drainage basis, hydrophysical approach to quantitative morphology, Bull. Geol. Soc. America, 56, 275-370, 1945.

[33] G.D.Knott, A Numbering System for Binary Trees, Comm. of A.C.M., 20, 2, 113-115, 1977.

[34] D.E.Knuth, The art of computer programming, Fundamental Algorithms, 1, Addison Wesley, 1973.

[35] C.C.Lee, D.T.Lee, C.K.Wong, Generating binary trees of bounded height, Acta Inf., 23, 529-544, 1986.

[36] C.L.Liu, Generation of k-ary trees, Rapport Inria. n°27, 1980 and Proceedings C.A.A.P.'80, 45-53, Lille 1973.

[37] A.Meir, J.W.Moon, On the altitude of nodes in random trees, Canad. J. of Math. 30, 997-1015, 1978.

[38] A.Nijenhuis, H.S.Wilf, Combinatorial Algorithms, second edition, Academic Press, N.Y., 1978.

[39] R.Otter, The Number of Trees, Ann. of Math., 49, 583-599, 1948.

[40] J.M.Pallo, Generating trees with n nodes and m leaves, Int. J. Comp. Math., 21, 133-144, 1987.

[41] J-G.Penaud, Arbres et animaux, mémoire d'habilitation à diriger les recherches, Université de Bordeaux 1, 1990.

[42] J-G.Penaud, Une nouvelle bijection pour les animaux dirigés, Actes du 22ième séminaire Lotharingien de combinatoire, Université de Strasbourg, 1989.

[43] J.L.Rémy, Un procédé itératif de dénombrement d'arbres binaires et son application à leur génération aléatoire, R.A.I.R.O. Informatique Théorique, 19, 2, 179-195, 1985.

[44] D.Rotem, On a correspondence between Binary Trees and a Certain Type of Permutation, Inf. Proc.Letters, 4, 1, 58-61, 1975.

[45] D.Rotem and Y.L.Varol, Generation of Binary Trees from Ballot Sequences, J.A.C.M., 25, 3, 396-404, 1978.

[46] M.P.Schützenberger, Context-free languages and pushdown automata, Information and Control, 6, 246-261, 1963.

[47] D.M.Silberger, Occurrences of the integer $\frac{(2n-2)!}{n!(n-1)!}$, Roczniki Polskiego Towarzystwa Math. I, vol. 13, 91-96, 1969.

[48] J.-M.Steyaert, Structure et complexité des algorithmes, Thèse de doctorat d'Etat, Université Paris 7, 1984.

[49] H.S. Stone, Parallel processing with the perfect shuffle, IEEE Transactions on Computers, vol.c-20, 2, 153-161, February 1971.

[50] A.N. Strahler, Hypsometric (area-altitude) analysis of erosional topology, Bull. Geol. Soc. Amer., 63, 1117-1142, 1952.

[51] X.G.Viennot, Combinatoire énumérative, notes de cours ENS Ulm, Paris, 1989.

[52] X.G.Viennot, Trees everywhere, Proc. C.A.A.P.'90, Lecture Notes in Computer Science 431, 18-41, 1990,

[53] J.S.Vitter, Optimum algorithms for two random sampling problems, Proc. F.O.C.S.-83, 65-75, 1983.

[54] J. Van Leeuwen (Editor), Handbook of Theoretical Computer Science, Volume 1, 886-894, 1990.

[55] H.S.Wilf, Combinatorial Algorithms : An Update, CBMS-NSF Regional Conference Series in Applied Mathematics, SIAM Pub.1989.

INDEX